MILLIMETER-WAVE WAVEGUIDES

Millimeter-Wave Waveguides

by

Dmitri Lioubtchenko
Helsinki University of Technology, Finland

Sergei Tretyakov
Helsinki University of Technology, Finland

and

Sergey Dudorov
Helsinki University of Technology, Finland

SPRINGER SCIENCE+BUSINESS MEDIA, LLC

A C.I.P. Catalogue record for this book is available from the Library of Congress.

Additional material to this book can be downloaded from http://extras.springer.com.

ISBN 978-1-4419-5378-0 ISBN 978-0-306-48724-8 (eBook)
DOI 10.1007/978-0-306-48724-8

Preface

This book is about the key elements of the millimeter-wave technology: various millimeter-wave waveguides, waveguide transitions, and devices.

The book can serve both as a tutorial presenting the basic theory and the main experimental techniques necessary for the work with millimeter-wave waveguides and millimeter-wave devices, as well as a monograph presenting new developments in this field. Examples show the use of millimeter-wave waveguides in the design of microwave devices and antennas.

Most of the new results and designs described in this book have been developed at the Radio Laboratory / SMARAD research unit of the Helsinki University of Technology, and the authors would like to thank the Laboratory director, Professor Antti Räisänen, and all the personnel for help and creative atmosphere.

Espoo, April 2003
Dmitri Lioubtchenko
Sergei Tretyakov
Sergey Dudorov

Table of Contents

Introduction

This book deals with the quickly developing millimeter and submillimeter wave technology. This part of the electromagnetic spectrum occupies the region from about 30 GHz to several hundreds of GHz, that is, between the microwave and infrared regions. At present, both microwave and optical regions are actively exploited, especially by the telecommunications industry. The millimeter and submillimeter frequencies were used in the past mainly by the military and in radioastronomy. It appears that now it is the time for exploiting these parts of the spectrum in a wider range of applications, as the technology gets mature. Indeed, potential applications are numerous due to many unique features of millimeter and submillimeter waves. To give a few examples, we can mention that submillimeter-wave imaging is rapidly becoming recognised as a new and effective diagnostic technique in medicine. This is because short electromagnetic waves can penetrate through many optically opaque materials. Similarly to the X-ray technique, submillimeter waves can provide an image, but without use of potentially harmful radiation. An application of a different nature could be the detection of chemical and biological threats. All materials emit millimeter and submillimeter waves, each having its own frequency pattern. These waves escape from, for example, envelopes and postage parcels, and can be detected by an appropriate sensitive receiver. Still another emerging application is in collision avoidance radars for cars. Since the wavelength is rather small, it becomes possible to design small but highly directive antennas for this purpose. At short distances, these radars are operational even under severe weather conditions. Also in telecommunications millimeter waves offer more bandwidth, and this technology is very promising especially for indoor cellular systems.

The millimeter-wave technology has much in common with the microwave technology on one side and with the optics on the other side. The same solutions for basic waveguides and resonators naturally can be used also in the millimeter frequency range. However, there are many peculiarities specific for the millimeter-wave devices that call for more investigations necessary for development and perfection of the new technology. In this book, we will consider in detail millimeter-wave waveguides and some devices built around millimeter-wave waveguide sections. Due to several reasons, in this frequency range different materials are used to fabricate dielectric waveguides (as compared to fiber optics, for example). New materials dictate a different cross section geometry that is compatible with the material fabrication requirements. Moreover, many materials used in these devices are anisotropic, which still complicates the analysis and design tasks.

We start the exposition with an introductory chapter (Chapter 1) which presents the general theory of electromagnetic waveguides. The results of this chapter are general in the sense that they apply for all frequency ranges. Special attention is given to the main principles of open waveguides and to analytical techniques that will be used throughout the book in the analysis of special types of millimeter-wave guides and devices. Also, the basic theory of microwave resonators is presented in Chapter 2 in a form convenient for understanding measurement techniques described in Chapter 6.

As we already mentioned, in many instances the materials from which millimeter-wave waveguides are made are anisotropic crystals. Chapter 3 presents the basic theory of electromagnetic waves in anisotropic crystals and in simple anisotropic waveguides. To design nonreciprocal devices like isolators, and electrically controllable devices such as phase shifters and scanning antennas, we use ferrite materials. Chapter 4 is devoted to these classes of waveguides. Microwave properties of ferrites are explained, along with the fundamental phenomena in ferrite-filled waveguides.

When studying any transmission line, at first we should understand its propagation properties, namely, estimate propagation constants and attenuation. The dielectric waveguide is an open transmission line, therefore one has to consider the field distribution both inside and outside the dielectric core, which complicates the problem. While for the circular cross section waveguide the solution can be found relatively easily, the rectangular cross section is more practical at millimeter waves, but more difficult for calculations. Many calculation methods are known from the literature, however, most of them are numerical, such as the finite element method, the finite difference method, the integral equation method, and so forth. The main disadvantages of them are the time and computing resources limitations and impossibility for an analytical analysis of the solution. From our point of view, the most appropriate methods for the case of rectangular dielectric waveguide are the so-called Marcatili's (and its variations) and Goell's methods. In Chapter 5, the classical Marcatili's and Goell's methods are described, and then adapted for the anisotropic dielectric case. For instance, this is useful for investigations of anisotropic Sapphire waveguides, when we have to find a suitable method for calculating its propagation constants against the frequency. Also, experimental results are presented.

For measuring dielectric properties of materials the open resonator technique is the most accurate at the millimeter-wave frequencies. Therefore, in Chapter 6 this method is described. At the end of that chapter, dielectric properties of some practical materials are summarized.

Dielectric waveguides are promising for millimeter-wave applications, however, most of the power sources have standard metal waveguides at the output. Also, metal waveguides have some advantages, such as mechanical strength, easy connectivity, etc. Therefore, when designing devices based on dielectric waveguides, one usually has to connect them to metal waveguides. Here, the problem of matching becomes important. In Chapter 7, the results of simulations and also experimental results are presented for matching a standard metal waveguide with a high-permittivity dielectric waveguide. One can see that the matching efficiency can be quite good in spite of that the aperture of the metal waveguide is several times larger than the cross section of the dielectric waveguide. Moreover, the transition structure is very simple and does not contain launching horns which are common for low-permittivity dielectric waveguides used at lower frequencies.

Finally, in Chapter 8 different waveguiding structures and devices based on them are reviewed. The most attractive application is probably the active dielectric waveguide, as it might be another principle of electromagnetic wave amplification. This field is not yet sufficiently investigated to produce practical devices. However, experimental results look promising.

The book combines tutorial material, analytical reviews, and original results of the authors. Some of these results have not been published before.

Chapter 1

General theory of waveguides

This introductory chapter presents the fundamental theory of regular waveguides, which is needed for understanding of advanced and novel structures. To understand measurements, we will need knowledge of microwave resonators, to be considered in Chapter 2. The exposition here is designed for solid understanding of the following chapters. In the literature, there are many good text books on waveguides, and for further reading we can recommend, for example, [1] (introductory material, modes in rectangular and circular waveguides), [2] (advanced theory), and [3] (excitation of open waveguides and other advanced topics). Analytical methods for solving various waveguide problems are described in [4–6]. The classical handbook [7] gives a useful collection of waveguide formulas.

Let us start from the definition of the regular waveguide:

Definition. *Regular or cylindrical waveguide is a system whose electromagnetic properties do not change along at least one straight line in space.*

This is a very general notion, covering, for example, free space, plane surfaces and interfaces, rods of an arbitrary cross section, all kinds of tubes and pipes, etc. Moreover, very many devices can be considered as sets of bulk elements connected via sections of regular waveguides. The general field solution can be found as a series of eigensolutions for regular waveguides that we will study next.

1.1 BASIC RELATIONS FOR REGULAR WAVEGUIDES

Let us introduce a coordinate system (for example, Cartesian or cylindrical) and direct the z-axis along the waveguide axis, that is, the direction in space along which the physical properties do not change (see the definition above).

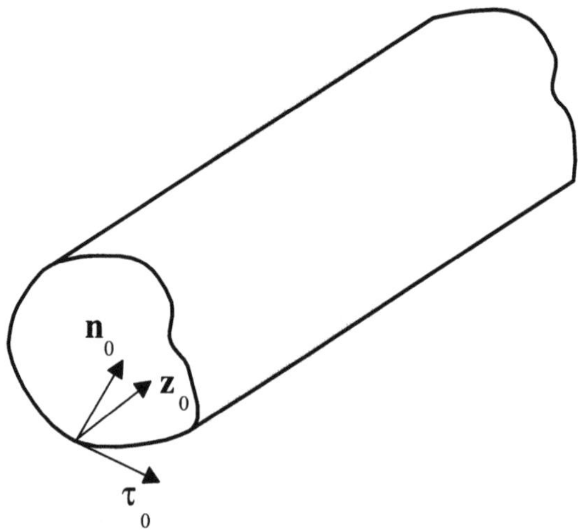

Figure 1.1. A generic cylindrical waveguide and three orthogonal coordinate unit vectors. Vector $\mathbf{z}_0$ is directed along the waveguide axis, vector $\mathbf{n}_0$ is orthogonal to the waveguide wall or an interface with a different material, and $\boldsymbol{\tau}_0$ is tangential to the wall and orthogonal to $\mathbf{z}_0$ (and orthogonal to $\mathbf{n}_0$ as well).

Here we suppose that the waveguide cross section is filled by a uniform isotropic medium described by scalar parameters ε and μ. Because there is only one physically defined special direction in space, that of the waveguide axis, it is of advantage to split the electromagnetic fields into the longitudinal (z) and transverse (t) components with respect to the this axis:

$$\mathbf{E} = \mathbf{E}_t + \mathbf{z}_0 E_z, \qquad \mathbf{H} = \mathbf{H}_t + \mathbf{z}_0 H_z \tag{1.1}$$

Here, $\mathbf{z}_0$ is the unit vector along the waveguide axis, and vectors $\mathbf{E}_t$ and $\mathbf{H}_t$ are orthogonal to $\mathbf{z}_0$. In the same way, we split the nabla operator:

$$\nabla = \nabla_t + \frac{\partial}{\partial z}\mathbf{z}_0 \tag{1.2}$$

With these notations, for time-harmonic fields [the time dependence is in the form $\exp(j\omega t)$], the Maxwell equations

$$\nabla \times \mathbf{E} = -j\omega\mu\mathbf{H}, \qquad \nabla \times \mathbf{H} = j\omega\varepsilon\mathbf{E} \tag{1.3}$$

for the fields in the waveguide take the form

$$\nabla_t \times \mathbf{E}_t + \nabla_t \times \mathbf{z}_0 E_z + \frac{\partial}{\partial z}\mathbf{z}_0 \times \mathbf{E}_t = -j\omega\mu(\mathbf{H}_t + \mathbf{z}_0 H_z) \tag{1.4}$$

$$\nabla_t \times \mathbf{H}_t + \nabla_t \times \mathbf{z}_0 H_z + \frac{\partial}{\partial z}\mathbf{z}_0 \times \mathbf{H}_t = j\omega\varepsilon(\mathbf{E}_t + \mathbf{z}_0 E_z) \tag{1.5}$$

Equating separately the longitudinal and transverse parts of the Maxwell equations, we have for the longitudinal part

$$\nabla_t \times \mathbf{E}_t = -j\omega\mu\mathbf{z}_0 H_z \tag{1.6}$$

$$\nabla_t \times \mathbf{H}_t = j\omega\varepsilon\mathbf{z}_0 E_z \tag{1.7}$$

The transverse parts of the Maxwell equations (1.4) and (1.5) read:

$$\nabla_t \times \mathbf{z}_0 E_z + \frac{\partial}{\partial z}\mathbf{z}_0 \times \mathbf{E}_t = -j\omega\mu\mathbf{H}_t \tag{1.8}$$

$$\nabla_t \times \mathbf{z}_0 H_z + \frac{\partial}{\partial z}\mathbf{z}_0 \times \mathbf{H}_t = j\omega\varepsilon\mathbf{E}_t \tag{1.9}$$

Equations (1.6) and (1.7) allow us to express the longitudinal field components in terms of the transverse derivatives of the transverse field components:

$$\mathbf{z}_0 H_z = -\frac{1}{j\omega\mu}\nabla_t \times \mathbf{E}_t, \qquad \mathbf{z}_0 E_z = \frac{1}{j\omega\varepsilon}\nabla_t \times \mathbf{H}_t \tag{1.10}$$

1.1.1 Vector transmission-line equations

Next, we can substitute the longitudinal fields from (1.10) into (1.8) and (1.9). The result is

$$\frac{\partial}{\partial z}\mathbf{z}_0 \times \mathbf{E}_t = -j\omega\mu\mathbf{H}_t - \frac{1}{j\omega\varepsilon}\nabla_t \times (\nabla_t \times \mathbf{H}_t) \tag{1.11}$$

$$\frac{\partial}{\partial z}\mathbf{z}_0 \times \mathbf{H}_t = j\omega\varepsilon\mathbf{E}_t + \frac{1}{j\omega\mu}\nabla_t \times (\nabla_t \times \mathbf{E}_t) \tag{1.12}$$

Cross-multiplying by $-\mathbf{z}_0$ from the left, this transforms to

$$\frac{\partial}{\partial z}\mathbf{E}_t = \left(j\omega\mu\overline{\overline{I}}_t + \frac{j}{\omega\varepsilon}\nabla_t\nabla_t\right)\cdot(\mathbf{z}_0 \times \mathbf{H}_t) \tag{1.13}$$

$$\frac{\partial}{\partial z}\mathbf{z}_0 \times \mathbf{H}_t = \left(j\omega\varepsilon\overline{\overline{I}}_t + \frac{j}{\omega\mu}\mathbf{z}_0 \times \nabla_t\, \mathbf{z}_0 \times \nabla_t \right) \cdot \mathbf{E}_t \tag{1.14}$$

Here, $\overline{\overline{I}}_t = \overline{\overline{I}} - \mathbf{z}_0\mathbf{z}_0$ is the two-dimensional unit dyadic in the transverse plane. These two equations can be called *vector transmission-line equations* because they connect "vector voltage" $\mathbf{E}_t$ and "vector current" $\mathbf{n} \times \mathbf{H}_t$ in the same way as the usual voltage and current are related in the conventional transmission-line equations. This deep analogy between waves in transmission lines, plane waves in free space, and waveguide modes comes from the fact that all these systems actually belong to the class of regular waveguides.

1.1.2 Longitudinal and transverse fields

Equations (1.8) and (1.9) allow us to express the transverse field components in terms of the longitudinal ones. The idea here is to work with the longitudinal components only, which is clearly much simpler. Consider wave solutions with $\exp(-jk_z z)$ dependence on the longitudinal coordinate,[1] where k_z is the propagation factor of a wave along the waveguide, and eliminate $\mathbf{E}_t$, substituting from (1.9) into (1.8):

$$\nabla_t \times \mathbf{z}_0 E_z - \frac{k_z}{\omega\varepsilon}\mathbf{z}_0 \times (\nabla_t \times \mathbf{z}_0 H_z - jk_z\mathbf{z}_0 \times \mathbf{H}_t) = -j\omega\mu\mathbf{H}_t \tag{1.15}$$

Using identities $\mathbf{z}_0 \times (\nabla_t \times \mathbf{z}_0 H_z) = \nabla_t H_z$, $\mathbf{z}_0 \times (\mathbf{z}_0 \times \mathbf{H}_t) = -\mathbf{H}_t$, and $\nabla_t \times \mathbf{z}_0 E_z = -\mathbf{z}_0 \times \nabla_t E_z$, we arrive at

$$\mathbf{H}_t = \frac{1}{k_z^2 - k^2}\left(jk_z\nabla_t H_z + j\omega\varepsilon\, \mathbf{z}_0 \times \nabla_t E_z\right) \tag{1.16}$$

where $k = \omega\sqrt{\varepsilon\mu}$ is the wave number in the filling material. In a similar way we obtain

$$\mathbf{E}_t = \frac{1}{k_z^2 - k^2}\left(jk_z\nabla_t E_z - j\omega\mu\, \mathbf{z}_0 \times \nabla_t H_z\right) \tag{1.17}$$

Thus, the transverse field components are expressed through two-dimensional gradients of the longitudinal fields.

We know that the total fields in the guide are subject to the Helmholtz equations

$$(\nabla^2 + k^2)\mathbf{E} = 0, \qquad (\nabla^2 + k^2)\mathbf{H} = 0 \tag{1.18}$$

[1] This is possible because we have a *waveguide*, that is, our system is *regular* along the waveguide axis z.

Every separate component of the fields satisfies the same equation. In particular, we can write

$$(\nabla^2 + k^2)E_z = 0, \qquad (\nabla^2 + k^2)H_z = 0 \tag{1.19}$$

For wave solutions, substituting $E_z \exp(-jk_z z)$, we get

$$\nabla_t^2 E_z + (k^2 - k_z^2)E_z = 0 \tag{1.20}$$

and also

$$\nabla_t^2 H_z + (k^2 - k_z^2)H_z = 0 \tag{1.21}$$

We can conclude that the problem can be tackled by solving only two *scalar* Helmholtz equations (1.20)–(1.21) for the longitudinal fields, which is a dramatic simplification compared to the full Maxwell equations. When the longitudinal fields are known, the transverse fields can be found from (1.16) and (1.17).

1.2 BOUNDARY CONDITIONS AND WAVEGUIDE MODES IN CLOSED GUIDES

The only assumption of the previous theory was about an isotropic and homogeneous filling of the waveguide. Now we assume that the waveguide is *closed* by an impenetrable for the electromagnetic fields boundary.

1.2.1 Dirichlet and Neumann boundary conditions

Consider waveguides with ideally conducting walls. The boundary condition on the wall is

$$\mathbf{n}_0 \times \mathbf{E}|_s = (\mathbf{n}_0 \times \mathbf{E}_t + \mathbf{n}_0 \times \mathbf{z}_0 E_z)|_s = 0 \tag{1.22}$$

Here $\mathbf{n}_0$ is the unit vector normal to the wall surface, Figure 1.1. The two vectors in the sum (1.22) are orthogonal, thus, both must vanish. We see that solutions to (1.20) must satisfy the Dirichlet boundary condition

$$E_z|_s = 0 \tag{1.23}$$

and also the following condition must be satisfied:

$$\mathbf{n}_0 \times \mathbf{E}_t|_s = 0 \tag{1.24}$$

(this is the other electric field component tangential to the wall surface). Now we are happy with (1.23) because we have the boundary condition for equation (1.20). But we also need a boundary condition for H_z that is still missing. Let us find it. This should follow from (1.24), so we take equation (1.17) for $\mathbf{E}_t$ at the wall surface and cross-multiply that by $\mathbf{n}_0$. The result must be zero at the surface of the ideally conducting wall:

$$\mathbf{n}_0 \times \mathbf{E}_t|_s = \frac{1}{k_z^2 - k^2} \left[jk_z \mathbf{n}_0 \times \nabla_t E_z - j\omega\mu \mathbf{n}_0 \times (\mathbf{z}_0 \times \nabla_t H_z) \right] \Big|_s = 0 \tag{1.25}$$

Now, $\mathbf{n}_0 \times \nabla_t E_z|_s = 0$ because $E_z|_s = 0$ and thus its gradient is parallel to $\mathbf{n}_0$. $\mathbf{n}_0 \times (\mathbf{z}_0 \times \nabla_t H_z) = \mathbf{z}_0(\mathbf{n}_0 \cdot \nabla_t H_z) = \mathbf{z}_0 \partial H_z / \partial n$. From this we see that (1.25) is equivalent to the Neumann condition

$$\left. \frac{\partial H_z}{\partial n} \right|_s = 0 \tag{1.26}$$

and this is the condition for H_z that we need.

1.2.2 TE and TM modes

From the previous analysis we conclude that the longitudinal field components E_z and H_z satisfy the scalar Helmholtz equation which we write as

$$\nabla_t^2 \Psi + k_c^2 \Psi = 0, \qquad \text{where} \quad k_c^2 = k^2 - k_z^2 \tag{1.27}$$

Here Ψ stands for either E_z or H_z. If the waveguide is closed and the walls are ideally conducting, the equations for E_z and H_z can be solved independently one from another, because also the boundary conditions split into two independent conditions: one for E_z [equation (1.23)], and one for H_z [equation (1.26)]. Physically this means that the eigenwaves in these waveguides can be classified into TE *(transverse electric)* waves with $H_z = 0$ and TM *(transverse magnetic)* waves with $E_z = 0$.[2]

In the Helmholtz equation (1.27) the unknown functions E_z and H_z depend on two variables: the transverse coordinates in the waveguide cross section. In case of ideally conducting walls we can consider separately the Dirichlet and Neumann problems for Ψ:

$$\Psi|_s = 0 \qquad \text{Dirichlet problem (TM modes)} \tag{1.28}$$

$$\left. \frac{\partial \Psi}{\partial n} \right|_s = 0 \qquad \text{Neumann problem (TE modes)} \tag{1.29}$$

---

[2]TE and TM modes can exist in some open waveguides as well, as will be explained later.

From the mathematics it is known that both these problems have infinite countable sets of real eigennumbers $k_{c1}, k_{c2}, \ldots\ k_{ci}, \ldots$, and $k_{ci} \to \infty$ for $i \to \infty$. For each of these eigennumbers there is a corresponding eigenfunction E_z or H_z. These eigensolutions are called *modes*. Thus, there are infinitely many modes with the propagation factors

$$k_{zi} = \pm\sqrt{k^2 - k_{ci}^2} \tag{1.30}$$

The two signs correspond to waves in the opposite directions along the waveguide. Of course, the eigennumbers k_{ci} and the corresponding eigenfunctions are in general different for TM and TE modes.

The following cases are possible:

- $k^2 > k_{ci}^2$. In this case k_z is real, which corresponds to propagating waves $\exp(\pm j k_z z)$.

- $k^2 < k_{ci}^2$. In this case k_z is purely imaginary, which corresponds to an exponentially decaying wave $\exp(\pm k_z z)$, called *evanescent mode* (the sign must be chosen so that the wave indeed decays away from the source).

- $k^2 = k_{ci}^2$. For every mode, this wavenumber $k = k_c$ is called the *critical wavenumber*, or *cut-off wavenumber*, because at this wavenumber (and at the corresponding frequency) the i-th mode becomes an evanescent mode.

Thus, the critical wavenumbers are equal to the corresponding eigenvalues $k_{ci} = \omega_{ci}\sqrt{\varepsilon\mu}$. For every given frequency ω only modes whose cut-off wavenumbers are smaller that $k = \omega\sqrt{\varepsilon\mu}$ can propagate. Thus, only a limited number of propagating modes exists at every finite frequency. When the frequency increases, more modes can propagate in a waveguide. Usually, the working regime is chosen to be the single-mode regime.

Along with the wavenumbers, the corresponding wavelengths are introduced: the usual plane-wave wavelength $\lambda = 2\pi/k$, wavelength of a waveguide mode $\lambda_g = 2\pi/k_z$, and the critical wavelength $\lambda_c = 2\pi/k_c$. In terms of these quantities, the definition $k_c^2 = k^2 - k_z^2$ reads

$$\frac{1}{\lambda_c^2} = \frac{1}{\lambda^2} - \frac{1}{\lambda_g^2} \tag{1.31}$$

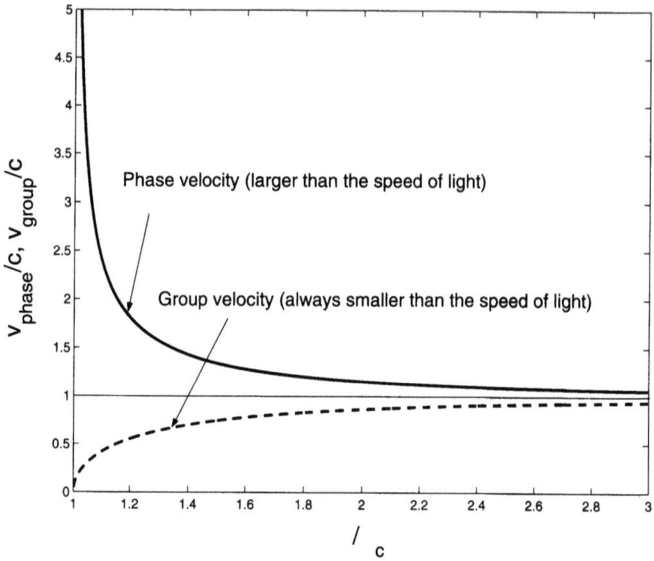

Figure 1.2. Phase and group velocities of a TE or TM mode in a waveguide. Note that the group velocity is always smaller than the speed of light. At the cut-off frequency the energy transport stops. At very high frequencies, waveguide modes behave as plane waves along z.

From here we have

$$\lambda_g = \frac{\lambda}{\sqrt{1 - \frac{\lambda^2}{\lambda_c^2}}} > \lambda \tag{1.32}$$

The phase velocity v is defined from $\omega t - k_z z = \text{const}$, thus

$$v = \frac{\omega}{k_z} = \frac{c}{\sqrt{1 - \frac{\omega_c^2}{\omega^2}}} > c \tag{1.33}$$

where $c = 1/\sqrt{\varepsilon\mu}$ is the speed of light in the filling material, and the critical frequency $\omega_c = k_c/\sqrt{\varepsilon\mu}$. The phase velocity in waveguides is greater than the speed of light, indeed $v \to \infty$ when $\omega \to \omega_c$.

The group velocity v_g is $d\omega/dk_z$. It is easier to calculate

$$\frac{dk_z}{d\omega} = \frac{1}{c}\frac{k}{\sqrt{k^2 - k_c^2}} \tag{1.34}$$

Thus,

$$v_g = c\sqrt{1 - \frac{k_c^2}{k^2}} = c\sqrt{1 - \frac{\omega_c^2}{\omega^2}} < c \tag{1.35}$$

This is smaller than the speed of light, and $v_g \to 0$ when $\omega \to \omega_c$: the energy transport stops at the cut-off frequency. Note, that $vv_g = c^2$. Both v and $v_g \to c$ for $\omega \to \infty$: at extremely large frequencies when the cross section size is very large compared to the wavelength in the filling material, waves in waveguides propagate like plane waves in the filling material.

1.2.3 Lossy waveguide walls. Hybrid modes

In reality, waveguide walls are not ideally conducting. For good conductors, one usually models surfaces introducing the impedance boundary condition

$$E_z \mathbf{z}_0 + E_\tau \boldsymbol{\tau}_0 = Z_s \mathbf{n}_0 \times \mathbf{H} \tag{1.36}$$

Here, Z_s equals to the wave impedance in the wall material, normally a metal:

$$Z_s = \sqrt{\frac{\mu_{\text{metal}}}{\varepsilon_{\text{metal}}}} \approx \frac{1+j}{\sqrt{2}} \sqrt{\frac{\omega\mu_0}{\sigma}} \tag{1.37}$$

(the approximation corresponds to the assumption that $\mu_{\text{metal}} \approx \mu_0$ and $\varepsilon_{\text{metal}} \approx -j\sigma/\omega$, where σ is the conductivity of metal.)

Obviously, here we cannot split this vector boundary condition into two scalar ones containing only electric or magnetic field, as was done for the ideally conducting walls. This means that the two main equations (1.20) and (1.21) are coupled via the boundary condition (1.36). All modes have all six non-zero field components, and these modes are called *hybrid*. However, if losses are small (meaning that $|Z_s|$ is very small), one can use the perturbation theory, assuming that the field distribution in the waveguide cross section is nearly the same as that in the same waveguide with an ideally conducting wall. In this sense we still call the modes TE and TM.

1.2.4 TEM mode

Previously we assumed that at least one of the longitudinal field components E_z and H_z is non-zero. To make the analysis complete let us consider the case when both $E_z \equiv 0$ and $H_z \equiv 0$. This solution is called TEM mode *(transverse electromagnetic)*, because both field vectors are transverse with respect to the waveguide axis. To have a non-trivial solution, both transverse fields must be non-zero. Looking at equations (1.16) and (1.17) we note that the experssions in brackets are zero. Thus, the only possible way to have non-zero left-hand sides is to nullify the denominator. This means that

$$k_z = \pm k \tag{1.38}$$

We conclude that TEM waves propagate (if propagate at all) with the speed of light in the material that fills the guide.

Equations for the transverse fields follow from (1.10). Because $E_z \equiv 0$ and $H_z \equiv 0$, we have simply

$$\nabla_t \times \mathbf{E}_t = 0, \qquad \nabla_t \times \mathbf{H}_t = 0 \tag{1.39}$$

These are the two-dimensional *static* equations. That is, the field distribution in every cross section coincides with the static distribution in the uniform infinite waveguide. Usually it is convenient to introduce a scalar potential Φ, so that $\mathbf{E}_t = -\nabla_t\Phi$, and look for solutions of equation $\nabla_t \cdot \mathbf{E}_t = 0$, which means the two-dimensional Laplace equation for the scalar potential

$$\nabla_t^2\Phi(x,y) = 0 \tag{1.40}$$

The boundary condition for ideally conducting walls is

$$\mathbf{E}_\tau|_s = 0, \qquad \text{that is} \quad \left.\frac{\partial\Phi}{\partial\tau}\right|_s = 0 \quad \text{or} \quad \Phi|_s = \text{const} \tag{1.41}$$

Non-trivial solutions are indeed possible, but only if we have at least two separate conductors. Single closed tubes do not support TEM waves.[3]

Finally we stress that a TEM wave, if such a solution exists, can propagate at all frequencies, from zero to infinity, without any dispersion (assuming that the filling material in non-dispersive). This is an exact solution to the Maxwell equations, there has been no quasi-static approximation made.

TEM solutions do not exist in waveguides with lossy walls. Indeed, splitting the magnetic field vector in the right-hand side of the boundary condition (1.36) into the longitudinal and the transverse parts, we get

$$\mathbf{E}_t + \mathbf{z}_0 E_z = Z_s(\mathbf{n}_0 \times \mathbf{H}_t + \mathbf{n}_0 \times \mathbf{z}_0 H_z) \tag{1.42}$$

We see that from the assumption $H_z|_s = 0$ and $E_z|_s = 0$ it follows that also $\mathbf{E}_t|_s = 0$ (as well as $\mathbf{H}_t|_s = 0$). This follows from the fact

[3]The only solution of the Laplace equation in a closed area with the condition $\Phi|_s = \text{const}$ on its boundary is $\Phi = \text{const}$ everywhere inside the area, which means that the electric field is identically zero.

that $\mathbf{n}_0 \times \mathbf{H}_t$ is a longitudinal vector, and $\mathbf{n}_0 \times \mathbf{z}_0 H_z$ is parallel to $\mathbf{E}_t$. No TEM solutions are possible also if the material filling the waveguide is inhomogeneous over the cross section, except some cases of special symmetry.

Waves with very small longitudinal field components and small dispersion can exist also in these situations, and they are called *quasi-TEM* waves. One typical example of a waveguide supporting a quasi-TEM wave is the microstrip line.

1.3 ORTHOGONALITY OF THE MODAL FIELDS

The usual definition of orthogonality of a set of real scalar functions Φ_i defined at $[a, b]$ is

$$\int_a^b \Phi_i(x)\Phi_j(x)\,dx = \begin{cases} 0 & \text{if } \; i \neq j \\ \text{const} \neq 0 & \text{if } \; i = j \end{cases} \tag{1.43}$$

For our purposes we introduce a similar definition for complex vector functions $\mathbf{A}_i$ (transverse fields in waveguides) of two variables (transverse coordinates) defined in a planar area S (waveguide cross section). The definition is

$$\int_S \mathbf{A}_i \cdot \mathbf{A}_j\,dS = \begin{cases} 0 & \text{if } \; i \neq j \\ \text{const} \neq 0 & \text{if } \; i = j \end{cases} \tag{1.44}$$

The following orthogonality relations hold:

$$\int_S \mathbf{E}_{ti} \cdot \mathbf{E}_{tj}\,dS = 0, \qquad \int_S \mathbf{H}_{ti} \cdot \mathbf{H}_{tj}\,dS = 0 \tag{1.45}$$

where the functions are the transverse fields of two different modes i and j. This is a very important fact allowing to look for waveguide solutions in terms of modal expansions.

1.3.1 The proof

Two TM modes

Consider two different TM modes. For TM modes $H_z = 0$, thus the relations for the transverse fields in terms of the longitudinal ones (1.17) simplify:

$$(k_{zi}^2 - k^2)\mathbf{E}_{ti} = jk_{zi}\nabla_t E_{zi} \tag{1.46}$$

$$(k_{zj}^2 - k^2)\mathbf{E}_{tj} = jk_{zj}\nabla_t E_{zj} \tag{1.47}$$

(we have written the relation for two modes i and j with different propagation constants k_{zi} and k_{zj}). Multiplying these equations and integrating them we get

$$(k_{zi}^2 - k^2)(k_{zj}^2 - k^2)\int_s \mathbf{E}_{ti}\cdot\mathbf{E}_{tj}\,dS = -k_{zi}k_{zj}\int_s \nabla_t E_{zi}\cdot\nabla_t E_{zj}\,dS \tag{1.48}$$

Next we make use of an identity valid for any differentiable scalar function ϕ and vector function $\mathbf{a}$ of two variables:

$$\nabla_t\cdot(\phi\mathbf{a}) = \mathbf{a}\cdot\nabla_t\phi + \phi\nabla_t\cdot\mathbf{a} \tag{1.49}$$

Let $\mathbf{a} = \nabla_t E_{zi}$ and $\phi = E_{zj}$. Then

$$\int_s \nabla_t E_{zi}\cdot\nabla_t E_{zj}\,dS = \int_s \nabla_t\cdot(E_{zj}\nabla_t E_{zi})\,dS - \int_s E_{zj}\nabla_t^2 E_{zi}\,dS \tag{1.50}$$

In the first integral we use the two-dimensional Gauss theorem, and in the second one we use the fact that E_{zi} satisfies the Helmholtz equation [so that $\nabla_t^2 E_{zi} = -(k^2 - k_{zi}^2)E_{zi} = -k_{ci}^2 E_{zi}$]. In doing so we obtain

$$\int_s \nabla_t E_{zi}\cdot\nabla_t E_{zj}\,dS = \int_L E_{zj}\frac{\partial E_{zi}}{\partial n}\,dl + (k^2 - k_{zi}^2)\int_s E_{zi}E_{zj}\,dS \tag{1.51}$$

The first integral is zero due to the boundary condition on the ideally conducting wall. The second is zero because these two functions are eigenfunctions of an eigenvalue problem for the Helmholtz equation (we assume a non-degenerate case here, so $k_{zi} \neq k_{zj}$). The last fact can be also proven directly. Indeed, we can apply the same transformation again but now setting $\mathbf{a} = \nabla_t E_{zj}$ and $\phi = E_{zi}$. In the same way we obtain

$$\int_s \nabla_t E_{zi}\cdot\nabla_t E_{zj}\,dS = (k^2 - k_{zj}^2)\int_s E_{zi}E_{zj}\,dS \tag{1.52}$$

The difference from (1.51) is that k_{zi} is replaced by k_{zj}. If $k_{zi} \neq k_{zj}$, we subtract this from (1.51) and finally find that

$$\int_s \nabla_t E_{zi}\cdot\nabla_t E_{zj}\,dS = 0 \tag{1.53}$$

Remembering (1.48) we see that indeed

$$\int_s \mathbf{E}_{ti}\cdot\mathbf{E}_{tj}\,dS = 0 \tag{1.54}$$

Note that the transverse magnetic fields are also orthogonal, which can be seen in a similar way. In case of degenerated modes, the set of longitudinal eigenfunctions can be first orthogonalized, and then the same conclusion for the transverse fields follows.

Two TE modes

This case can be considered in a similar way. We write for the transverse fields

$$(k_{zi}^2 - k^2)\mathbf{H}_{ti} = jk_{zi}\nabla_t H_{zi} \tag{1.55}$$

$$(k_{zj}^2 - k^2)\mathbf{H}_{tj} = jk_{zj}\nabla_t H_{zj} \tag{1.56}$$

The following derivation goes along the same line. The difference is that the line integral vanishes because of the Neumann condition imposed on the longitudinal magnetic field.

Modes of different types

Consider a TM mode whose transverse electric field is expressed through its longitudinal electric field component E_{zi} as

$$(k_{zi} - k^2)\mathbf{E}_{ti} = jk_{zi}\nabla_t E_{zi} \tag{1.57}$$

For a TE mode we have

$$(k_{zj}^2 - k^2)\mathbf{E}_{tj} = -j\omega\mu\mathbf{z}_0 \times \nabla_t H_{zj} \tag{1.58}$$

Multiplying these two equations and integrating over the waveguide cross section we get

$$(k_{zi}^2 - k^2)(k_{zj}^2 - k^2)\int_s \mathbf{E}_{ti}\cdot\mathbf{E}_{tj}\,dS = k_{zi}\omega\mu\int_s \nabla_t E_{zi}\cdot(\mathbf{z}_0\times\nabla_t H_{zj})\,dS \tag{1.59}$$

Let us invoke (1.49) again, now with $\mathbf{a} = \mathbf{z}_0 \times \nabla_t H_{zj}$ and $\phi = E_{zi}$. This transforms the right-hand-side integral as

$$\int_s \nabla_t E_{zi}\cdot(\mathbf{z}_0\times\nabla_t H_{zj})\,dS$$

$$= \int_s \nabla_t\cdot[E_{zi}(\mathbf{z}_0\times\nabla_t H_{zj})]\,dS - \int_s E_{zi}\nabla_t\cdot(\mathbf{z}_0\times\nabla_t H_{zj})\,dS \tag{1.60}$$

Here the first integral can be transformed into a line integral around the waveguide which vanishes because $E_{zi} = 0$ on the waveguide wall. In the second integral, $\nabla_t \cdot (\mathbf{z}_0 \times \nabla_t H_{zj}) = -\mathbf{z}_0 \cdot (\nabla_t \times \nabla_t H_{zj}) = 0$. Thus, we have proved that

$$\int_s \mathbf{E}_{ti} \cdot \mathbf{E}_{tj}\, dS = 0 \tag{1.61}$$

Note that in this case the result is true for arbitrary $k_{zi} \neq k$ and $k_{zj} \neq k$ (possibly also $k_{zi} = k_{zj}$).

1.4 FUNDAMENTAL PROPERTIES OF OPEN WAVEGUIDES

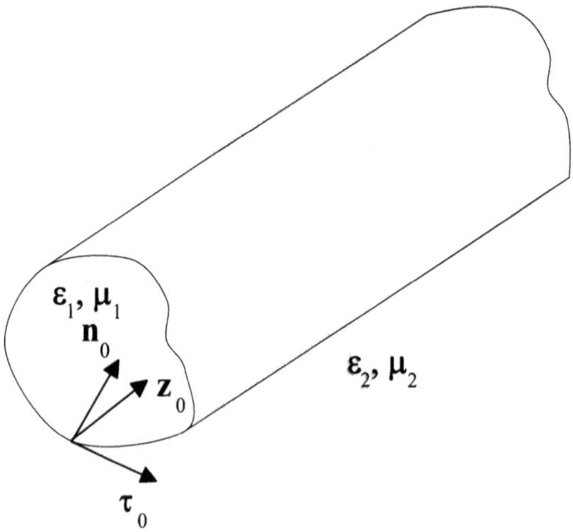

Figure 1.3. A generic open cylindrical waveguide. A cylinder with an arbitrary cross section made of a homogeneous material with parameters ε_1, μ_1 is located in another homogeneous and isotropic medium with parameters ε_2, μ_2.

In the millimeter-wave techniques, open waveguides are used very often, due to their smaller losses and other attractive features. Open waveguides (Figure 1.3) satisfy the general definition of the waveguide, so the same general approach can be applied also here. For the wave solutions $E_z \exp(-jk_z z)$ and $H_z \exp(-jk_z z)$ the longitudinal field components E_z and H_z satisfy the Helmholtz equations (1.20), (1.21); only the values of the wavenumber k are different in different media:

$$\nabla_t^2 E_z + (k_{1,2}^2 - k_z^2) E_z = 0 \tag{1.62}$$

and

$$\nabla_t^2 H_z + (k_{1,2}^2 - k_z^2) H_z = 0 \tag{1.63}$$

where $k_{1,2} = \omega\sqrt{\varepsilon_{1,2}\mu_{1,2}}$. To solve them, we need boundary conditions for the longitudinal fields on the interface between the two media. We will derive the general relations for arbitrary cross section waveguides here, which we will further need to discuss general properties of open waveguides.

1.4.1 Boundary conditions for open waveguides

The usual boundary conditions on an interface of two different media (continuity of the tangential field components) read

$$E_{z1}|_s = E_{z2}|_s, \qquad H_{z1}|_s = H_{z2}|_s \tag{1.64}$$

$$\boldsymbol{\tau}_0 \cdot \mathbf{E}_{t1}|_s = \boldsymbol{\tau}_0 \cdot \mathbf{E}_{t2}|_s, \qquad \boldsymbol{\tau}_0 \cdot \mathbf{H}_{t1}|_s = \boldsymbol{\tau}_0 \cdot \mathbf{H}_{t2}|_s \tag{1.65}$$

The index t denotes, as before, the transverse field components. $\boldsymbol{\tau}_0$ is the unit vector tangential to the line around the waveguide cross-section boundary:

$$\boldsymbol{\tau}_0 = \mathbf{z}_0 \times \mathbf{n}_0 \tag{1.66}$$

(see Figure 1.3).

Our goal will be to express (1.65) in terms of the longitudinal field components only, so that we will be able to solve (1.62) and (1.63). Let us substitute the known expressions for the transverse fields in terms of the longitudinal ones (1.16) and (1.17) into the last relations. The condition for $\mathbf{E}_t$ in (1.65) reads:

$$\frac{1}{k_z^2 - k_1^2}\left[jk_z\boldsymbol{\tau}_0 \cdot \nabla_t E_{z1} - j\omega\mu_1\boldsymbol{\tau}_0 \cdot (\mathbf{z}_0 \times \nabla_t H_{z1})\right]\Big|_s$$

$$= \frac{1}{k_z^2 - k_2^2}\left[jk_z\boldsymbol{\tau}_0 \cdot \nabla_t E_{z2} - j\omega\mu_2\boldsymbol{\tau}_0 \cdot (\mathbf{z}_0 \times \nabla_t H_{z2})\right]\Big|_s \tag{1.67}$$

and similarly for $\mathbf{H}_t$. This can be simplified because

$$\boldsymbol{\tau}_0 \cdot \nabla_t E_z = \frac{\partial E_z}{\partial \tau} \tag{1.68}$$

and

$$\boldsymbol{\tau}_0 \cdot (\mathbf{z}_0 \times \nabla_t H_z) = (\boldsymbol{\tau}_0 \times \mathbf{z}_0) \cdot \nabla_t H_z = \mathbf{n}_0 \cdot \nabla H_z = \frac{\partial H_z}{\partial n} \tag{1.69}$$

Finally,

$$\frac{1}{k_z^2 - k_1^2}\left(jk_z\frac{\partial E_{z1}}{\partial \tau} - j\omega\mu_1\frac{\partial H_{z1}}{\partial n}\right)\Big|_s$$

$$= \frac{1}{k_z^2 - k_2^2} \left(jk_z \frac{\partial E_{z2}}{\partial \tau} - j\omega\mu_2 \frac{\partial H_{z2}}{\partial n} \right) \bigg|_s \tag{1.70}$$

In the same way from the second relation of (1.65) we get

$$\frac{1}{k_z^2 - k_1^2} \left(jk_z \frac{\partial H_{z1}}{\partial \tau} + j\omega\varepsilon_1 \frac{\partial E_{z1}}{\partial n} \right) \bigg|_s$$

$$= \frac{1}{k_z^2 - k_2^2} \left(jk_z \frac{\partial H_{z2}}{\partial \tau} + j\omega\varepsilon_2 \frac{\partial E_{z2}}{\partial n} \right) \bigg|_s \tag{1.71}$$

Relations (1.70) and (1.71) together with (1.64) are the boundary conditions for the longitudinal fields E_z and H_z.

In general, the boundary conditions couple the electric and magnetic fields, thus there are no TE or TM modes, and the solutions are *hybrid* modes. An important exception is when the symmetry allows to have no dependence on τ. For example, for a dielectric fiber of the circular cross section there exists a class of solutions with no dependence on the polar angle ϕ. In that case $\frac{\partial E_z}{\partial \tau} = 0$ and $\frac{\partial H_z}{\partial \tau} = 0$, so that the boundary conditions (1.70) and (1.71) simplify:

$$\frac{\mu_1}{k_z^2 - k_1^2} \frac{\partial H_{z1}}{\partial n} \bigg|_s = \frac{\mu_2}{k_z^2 - k_2^2} \frac{\partial H_{z2}}{\partial n} \bigg|_s, \qquad H_{z1}|_s = H_{z2}|_s \tag{1.72}$$

$$\frac{\varepsilon_1}{k_z^2 - k_1^2} \frac{\partial E_{z1}}{\partial n} \bigg|_s = \frac{\varepsilon_2}{k_z^2 - k_2^2} \frac{\partial E_{z2}}{\partial n} \bigg|_s, \qquad E_{z1}|_s = E_{z2}|_s \tag{1.73}$$

As we see, the boundary conditions split into two sets of conditions for only electric or only magnetic fields. Thus, rotationally symmetric modes are TE and TM modes. Another important case when TE and TM solutions are allowed is the planar dielectric waveguide.

1.4.2 Eigenwaves in planar dielectric waveguides

Here we consider the main properties of open waveguides using a simple example of electromagnetic waves along a dielectric slab, see Figure 1.4.

To study the eigenwaves we will make use of the general theory of waveguides. That is, we will solve the Helmholtz equations for the longitudinal fields. Consider eigensolutions for fields with no dependence on one of the transverse coordinates (x), then equations (1.62) and (1.63) simplify to

$$\frac{\partial^2}{\partial y^2} E_z + (k^2 - k_z^2) E_z = 0, \qquad \frac{\partial^2}{\partial y^2} H_z + (k^2 - k_z^2) H_z = 0 \tag{1.74}$$

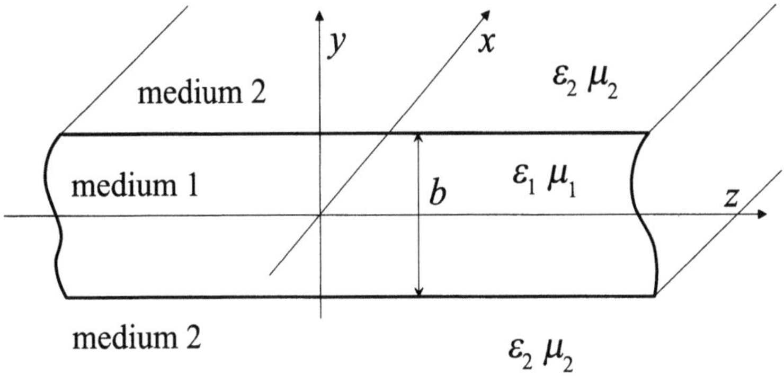

Figure 1.4. Plane dielectric waveguide. Waves travel along z, no field dependence on y.

For shortness, let us write only equations for TM modes, that is, for E_z. In equations (1.74) the wavenumber $k = \omega\sqrt{\varepsilon\mu}$ is different in the core and in the cladding (media 1 and 2 in Figure 1.4), which means that we have to write equations for these regions separately:

$$\frac{\partial^2}{\partial y^2}E_{z1} + (k_1^2 - k_z^2)E_{z1} = 0, \qquad \frac{\partial^2}{\partial y^2}E_{z2} + (k_2^2 - k_z^2)E_{z2} = 0 \quad (1.75)$$

where $k_{1,2} = \omega\sqrt{\varepsilon_{1,2}\mu_{1,2}}$. The propagation factor k_z is the same in both regions, as dictated by the boundary conditions on the interfaces.

From the general analysis we conclude that in our planar waveguide the solutions are TE and TM modes, and the boundary conditions for the TM solutions read

$$\frac{\varepsilon_1}{k_z^2 - k_1^2}\frac{\partial E_{z1}}{\partial y}\bigg|_s = \frac{\varepsilon_2}{k_z^2 - k_2^2}\frac{\partial E_{z2}}{\partial y}\bigg|_s, \qquad E_{z1}|_s = E_{z2}|_s \quad (1.76)$$

Looking for guided-wave solutions, it is convenient to write the general solution of (1.75) as

$$E_{z1} = A\sin(py) + B\cos(py), \qquad E_{z2} = Ce^{-q|y|} \quad (1.77)$$

Here we have denoted $k_1^2 - k_z^2 = p^2$ and $k_2^2 - k_z^2 = -q^2$. Such notations are convenient because for guided wave solutions both p and q are real numbers. Subtracting these two definitions we get

$$p^2 + q^2 = k_1^2 - k_2^2 = \omega^2(\varepsilon_1\mu_1 - \varepsilon_2\mu_2) \quad (1.78)$$

(for generality we assume that also the permeabilities of the two media can be different).

Next we note that it is possible to consider even and odd field distributions separately. For even modes at $y > 0$

$$E_{z1} = B\cos(py), \qquad E_{z2} = Ce^{-qy} \tag{1.79}$$

and the boundary conditions read

$$E_{z1}|_{y=b/2} = E_{z2}|_{y=b/2}, \qquad -\frac{\varepsilon_1}{p^2}\frac{\partial E_{z1}}{\partial y}\bigg|_{y=b/2} = \frac{\varepsilon_2}{q^2}\frac{\partial E_{z2}}{\partial y}\bigg|_{y=b/2} \tag{1.80}$$

Substituting (1.79) and equating the determinant to zero we arrive at the eigenvalue equation

$$\frac{qb}{2} = -\frac{\varepsilon_2}{\varepsilon_1}\frac{pb}{2}\cot\frac{pb}{2} \tag{1.81}$$

For the odd modes, substituting $E_{z1} = A\sin(px)$ we get

$$\frac{qb}{2} = \frac{\varepsilon_2}{\varepsilon_1}\frac{pb}{2}\tan\frac{pb}{2} \tag{1.82}$$

Equations for the TE modes can be found from here by replacing $\varepsilon_{1,2}$ by $\mu_{1,2}$.

The solutions can be found graphically, as shown in Figure 1.5. Here, the normalized transverse decay factor $qb/2$ is plotted as a function of the normalized transverse wave number $pb/2$, according to (1.81) and (1.82). These two parameters are also connected by (1.78), and the corresponding circle is also plotted here. The solutions are found at the crossings of these curves. The radius of the circle $R = \omega b\sqrt{\varepsilon_1\mu_1 - \varepsilon_2\mu_2}/2$ is called the *normalized frequency.* When this parameter is smaller than $\pi/2$, there is only one propagating mode *(the single-mode regime).* Typical for open waveguides is that this parameter depends on the material parameter contrast between the core and cladding: for dielectric waveguides, on the difference of the permittivities $\varepsilon_1 - \varepsilon_2$. If this difference is small, the single-mode regime can exist in a wide frequency band even if the cross section size is much larger than the wavelength. Such fibers find extremely important applications in telecommunications. They are called *weakly-guiding fibers.*

At the cut-off frequency of a mode the transverse decay factor q equals zero, see Figure 1.5. This means that the corresponding wave is not anymore guided by the slab. Note that in contrast to closed waveguides, the wave still does propagate along the slab. The propagation factor

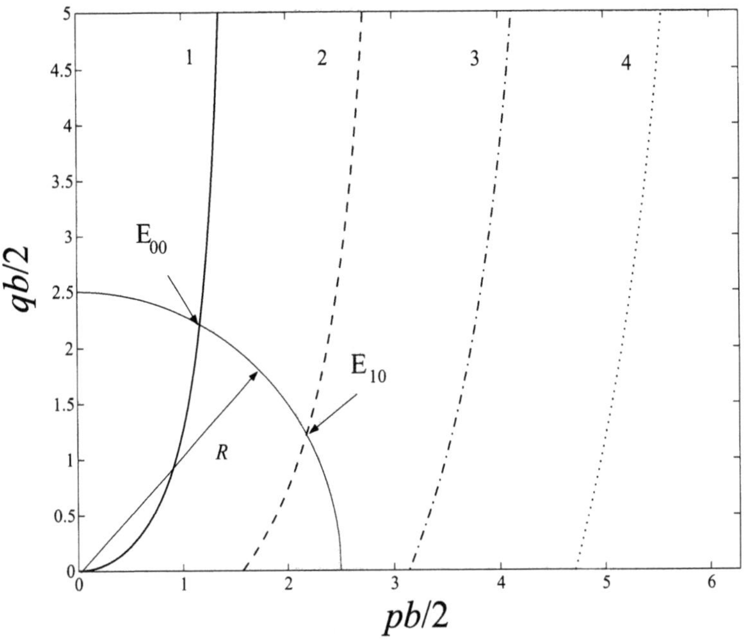

Figure 1.5. Graphical solution of the dispersion equations. Radius $R = \omega b\sqrt{\varepsilon_1\mu_1 - \varepsilon_2\mu_2}/2$. Curves 1 and 3 correspond to equation (1.81). Curves 2 and 4 correspond to (1.82).

of a wave at cut-off just equals that in the outer space: $k_z^2 = k_2^2 + q^2 = k_2^2$ if $q = 0$. In closed metal waveguides $k_z = 0$ at cut-off. The fundamental mode has no cut-off and can propagate at all frequencies. However, at low frequencies the transverse decay factor q becomes very small, which makes the waveguide impractical. If we now consider the behavior of a certain mode at very high frequencies, then we observe that in this limit p remains finite, although q tends to infinity. Obviously, $k_z \to k_1$, and the field is concentrated in the slab.

Figure 1.6 presents an example of typical dispersion curves for a millimeter-wave waveguide (silicon, the relative permittivity $\varepsilon_r = 11.56$). Near the cut-off of every mode the curve is close to the light line for free space, meaning a weak field concentration inside the slab. At high frequencies the curves asymptotically approach the light line for the filling material, since the field is more and more concentrated inside the slab.

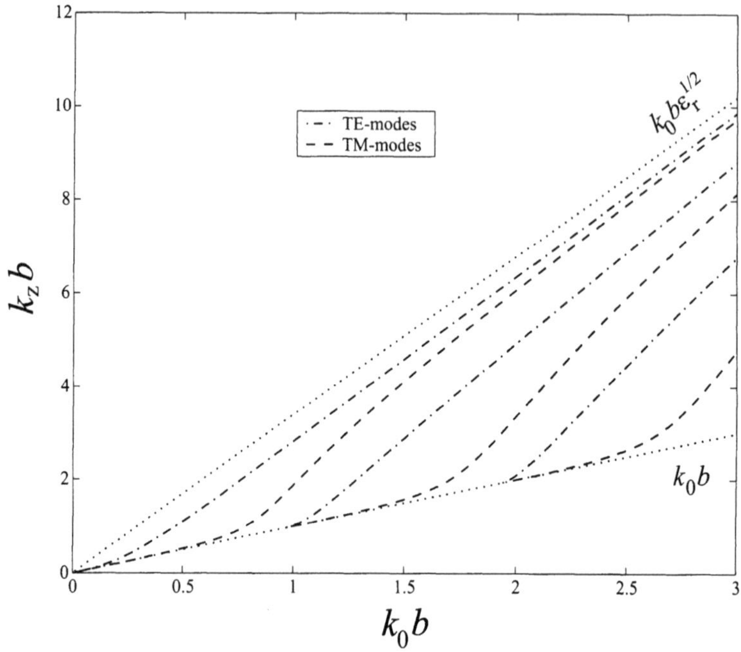

Figure 1.6. Typical dispersion curves for a planar dielectric waveguide. Dotted lines show the dispersion lines for free space ($k_0 b$) and the unbounded material filling the slab ($k_0 b\sqrt{\varepsilon_r}$).

1.5 INHOMOGENEITIES IN WAVEGUIDES

For understanding various inhomogeneities in waveguides, an analogy with the transmission-line theory is very useful.

1.5.1 Transmission-line theory applied to waveguides

Consider a single-mode waveguide. The general solution for the transverse fields is the sum of two waves traveling in the opposite directions along the guide axis:

$$\mathbf{E}_t(x, y, z) = \mathbf{E}_t^+(x, y)e^{-jk_z z} + \mathbf{E}_t^-(x, y)e^{jk_z z} \tag{1.83}$$

$$\mathbf{H}_t(x, y, z) = \mathbf{H}_t^+(x, y)e^{-jk_z z} + \mathbf{H}_t^-(x, y)e^{jk_z z} \tag{1.84}$$

Functions $\mathbf{E}_t^+(x, y)$ and $\mathbf{E}_t^-(x, y)$ are actually the same, and differ only by an amplitude coefficient, because they are solutions to the same eigenvalue problem. This is correct also for $\mathbf{H}_t^+(x, y)$ and $\mathbf{H}_t^-(x, y)$. Indeed, considering TE or TM waves, we see from equations (1.13) and (1.14) that $\mathbf{H}_t^{\pm}$ are proportional to $\mathbf{z}_0 \times \mathbf{E}_t^{\pm}$ (see more in the next section). Thus, we can rewrite (1.83)–(1.84) as

$$\mathbf{E}_t(x,y,z) = \mathbf{E}_t^{+}(x,y)e^{-jk_z z}\left[1+\Gamma(z)\right] \tag{1.85}$$

and

$$\mathbf{H}_t(x,y,z) = \mathbf{H}_t^{+}(x,y)e^{-jk_z z}\left[1-\Gamma(z)\right] \tag{1.86}$$

where $\Gamma(z) = \Gamma(0)e^{2jk_z z}$ is the *reflection coefficient.*

Next, we *define* the normalized impedance in the same way as in the usual transmission-line theory:

$$Z(z) = \frac{1+\Gamma(z)}{1-\Gamma(z)} \tag{1.87}$$

Then, of course,

$$\Gamma(z) = \frac{Z(z)-1}{Z(z)+1} \tag{1.88}$$

These parameters are well defined and can be directly measured. Let us stress that the "denormalization" of the normalized impedance is not possible. No voltage and current in the usual sense can be introduced. If needed, equivalent voltage and currents can be defined in terms of the transmitted power P_{inc} (which is a measurable quantity, of course):

$$U_{\text{inc}}(z) = \sqrt{P_{\text{inc}}}\, e^{j\arg\{E_t^{+}(z)\}} \tag{1.89}$$

The dimension of this quantity is not [V], but [$\sqrt{\text{W}}$]. We cannot define the wave impedance if the fields are not purely transverse. To overcome this trouble, we can *define* the current as simply

$$I_{\text{inc}} = U_{\text{inc}}, \qquad I_{\text{ref}} = -U_{\text{ref}} \tag{1.90}$$

and use the standard transmission-line theory machinery.

1.5.2 Equivalent circuits for basic inhomogeneities

We know that waveguides with a single-mode regime can be considered as transmission lines, and the transmission-line theory can be used for practical calculations. Here we will describe the parameters of various inhomogeneities in waveguide circuits. We will start from a simple connection of two waveguides of the same cross section filled by different

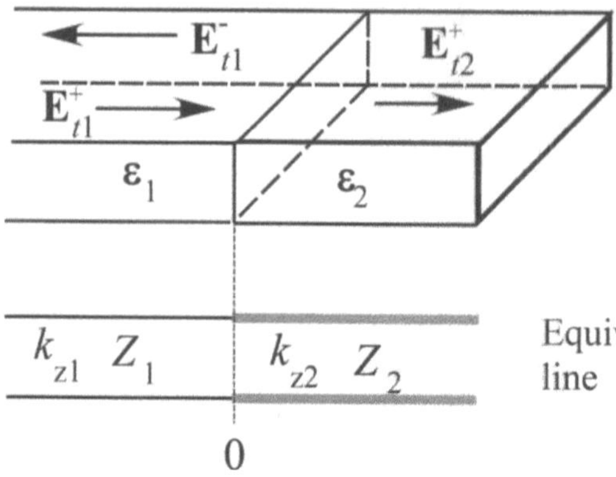

Figure 1.7. Two connected rectangular waveguides filled by different dielectrics.

materials (Figure 1.7[4]) and illustrate how the transmission-line parameters are introduced.

If both the waveguides work in the single-mode regime, this problem can be solved exactly. It is one of very rare cases when no higher-order modes are excited at an inhomogeneity. This is because the fundamental mode fields have the same distribution in the transverse plane in both waveguides. Let us consider, for example, a TE mode. Then the transverse components of the electric and magnetic fields can be expressed in terms of the longitudinal magnetic field component H_z as (1.16)–(1.17):

$$\mathbf{H}_t = \frac{jk_z}{k_z^2 - k^2}\nabla_t H_z \tag{1.91}$$

$$\mathbf{E}_t = \frac{-j\omega\mu}{k_z^2 - k^2}\mathbf{z}_0 \times \nabla_t H_z \tag{1.92}$$

Here $k_z = \sqrt{k^2 - k_c^2}$ is the propagation factor, and $\mathbf{z}_0$ is the unit vector along the waveguide. Combining these two equations we see that

$$\mathbf{H}_t = \frac{k_z}{\omega\mu}\mathbf{z}_0 \times \mathbf{E}_t, \qquad \text{or} \quad \mathbf{E}_t = -\frac{\omega\mu}{k_z}\mathbf{z}_0 \times \mathbf{H}_t \tag{1.93}$$

These is an extremely important result, because it shows that $\mathbf{E}_t$ and $-\mathbf{z}_0 \times \mathbf{H}_t$ have the same dependence on the transverse coordinates: they differ by a constant coefficient. This allows to interpret $\mathbf{E}_t$ as voltage

[4]The figures for this section have been drawn by A.S. Cherepanov, St. Petersburg State Technical University, Russia.

and $-\mathbf{z}_0 \times \mathbf{H}_t$ as current. The coefficient $\omega\mu/k_z$ plays the role of the wave impedance:[5]

$$\mathbf{E}_t = -\frac{\omega\mu}{k_z}\mathbf{z}_0 \times \mathbf{H}_t = Z_0(-\mathbf{z}_0 \times \mathbf{H}_t) \tag{1.94}$$

Now we can write the boundary conditions of continuity of the transverse fields in the plane of the junction:

$$\mathbf{E}_{t1}^+(1+\Gamma) = \mathbf{E}_{t2}^+ \tag{1.95}$$

$$\mathbf{H}_{t1}^+(1-\Gamma) = \mathbf{H}_{t2}^+ \tag{1.96}$$

The minus sign in the second equation comes about because the coefficient which connects the transverse electric and magnetic fields in (1.93) reverses sign when the propagation direction is reversed (that is, when the propagation constant k_z changes sign). Using (1.93), equation (1.96) transforms to

$$\frac{k_{z1}}{\omega\mu_1}\mathbf{E}_{t1}^+(1-\Gamma) = \frac{k_{z2}}{\omega\mu_2}\mathbf{E}_{t2}^+ \tag{1.97}$$

Indices 1 and 2 refer to the first and the second waveguide, respectively. The solution of (1.95) and (1.97) for the reflection coefficient is

$$\Gamma = \frac{\frac{k_{z1}}{\omega\mu_1} - \frac{k_{z2}}{\omega\mu_2}}{\frac{k_{z1}}{\omega\mu_1} + \frac{k_{z2}}{\omega\mu_2}} = \frac{\frac{1}{Z_1} - \frac{1}{Z_2}}{\frac{1}{Z_1} + \frac{1}{Z_2}} = \frac{1 - \frac{Z_1}{Z_2}}{1 + \frac{Z_1}{Z_2}} \tag{1.98}$$

Here we have used our interpretation in terms of the wave impedances $Z_{1,2}$. In fact, only the last expression which contains the *ratio* of Z_1 and Z_2 always has a well-defined physical meaning. Since there is no unique definition for the voltage and current, the wave impedance as such is not well defined. But the *ratio* of the wave impedances of two connected waveguides has a clear physical meaning and can be directly measured. Indeed, we can always measure the reflection coefficient Γ, and from (1.98) we get

$$\frac{Z_1}{Z_2} = \frac{1-\Gamma}{1+\Gamma} \tag{1.99}$$

The knowledge of this value (together with the propagation factor k_z) is enough to use the transmission-line theory, including the Smith chart.

---

[5]If the transverse dimensions stretch to infinity (or the frequency tends to infinity), then $k_z \to k$, and $\omega\mu/k_z \to \sqrt{\mu/\varepsilon}$. We recognize the plane-wave impedance.

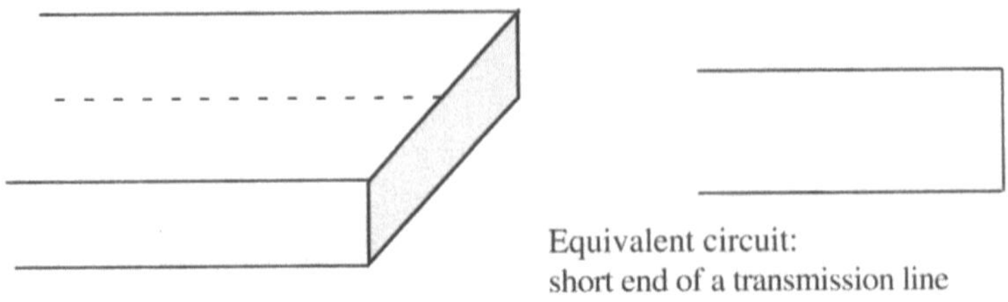

Figure 1.8. Short end of a waveguide section and its equivalent circuit.

Consider a section of a closed waveguide section which is terminated at the end by a metal wall orthogonal to the waveguide axis (Figure 1.8). The boundary condition at the end is

$$\mathbf{E}_t = \mathbf{E}_t^+ + \mathbf{E}_t^- = \mathbf{E}_t^+(1+\Gamma) = 0 \tag{1.100}$$

Here we assume that there is only one waveguide mode with the transverse electric field $\mathbf{E}_t$. As before, indices $\pm$ correspond to the waves traveling in the opposite directions. By the definition of the reflection coefficient, in this case $\Gamma = -1$. The same reflection coefficient has a short-circuited transmission line, thus, the equivalent circuit looks as in Figure 1.8.

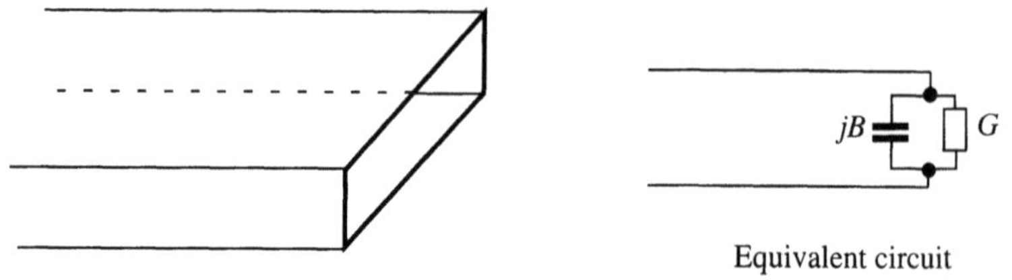

Figure 1.9. Open end of a waveguide section and its equivalent circuit.

Consider now an open end of a waveguide. In contrast to open ends of transmission lines which reflect all power back into the line ($\Gamma = 1$), an open end of a waveguide is an antenna, because its transverse dimensions are comparable to the wavelength. Thus, the equivalent circuit should contain a resistor equal to the radiation resistance of this antenna, see Figure 1.9. In addition, there is a reactive component (in metal waveguides, it can be approximated as a capacitance), because near the inhomogeneity there exist higher-order modes. In the fields of higher-order modes there is some stored energy. To find the parameters

G and B, the corresponding electromagnetic problem must be solved analytically or numerically. The Wiener-Hopf technique can be used for the analytical solution (e.g., [4]). In practice, it is usually much easier to measure the parameters directly, measuring the amplitude and phase of the reflection coefficient. For metal rectangular waveguides, usually $|\Gamma| \approx 1/3 \dots 1/4$. It is important to understand that the equivalent circuit is approximate also in the sense that the equivalent parameters are *not quite* capacitances, resistances, etc. The exact solution in this case, for example, leads to *frequency-dependent* capacitance and resistance.

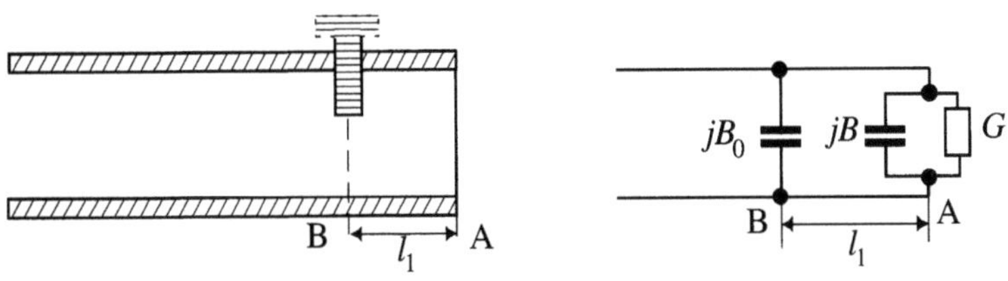

Figure 1.10. Matching of an open end of a waveguide.

This "antenna" can be matched by a small inhomogeneity, such as a simple screw, that is equivalent to a capacitive or inductive stub in transmission lines, see Figures 1.10 and 1.11. As soon as the reflection coefficient from the end has been measured and the equivalent parameters found, the distance to the screw l_1 can be determined using the Smith chart, and the value of the reactance adjusted experimentally varying the depth of the screw. Any inhomogeneity can be used for matching, see Figure 1.11. In this particular "experiment", the reflection was reduced by about 10 dB in the whole working frequency band of the waveguide.

Thin metal diaphragms with vertical slots (Figure 1.12) behave as inductive inclusions. This is because vertical electric currents flow on the inclusion, and there is some extra energy stored in the magnetic field of these currents. The equivalent inductance can be of course directly determined through the measured reflection coefficient. Theoretical calculation is somewhat involved, since many higher-order modes excited at the inhomogeneity have to be taken into account. Let us give an outline of a possible approach [6]. Assume that in a rectangular waveguide (Figure 1.12) only the fundamental mode is propagating. The x-component of the electric field of the incident wave is

$$E_x = \sin \frac{\pi y}{a} e^{-jk_z z} \tag{1.101}$$

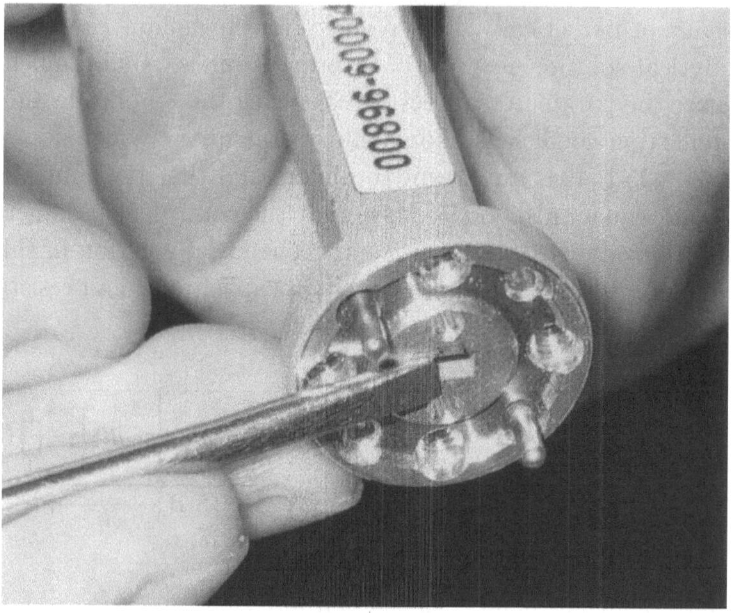

Figure 1.11. One possible way to decrease reflections from an open end of a waveguide.

Let the diaphragm be located at $z = 0$. The total electric field at the side of the source is the sum of the incident wave, reflected wave (of the main mode) and all higher-order modes with fields independent from x:

$$E_x(y, z) = \sin \frac{\pi y}{a} \left(e^{-jk_z z} + Re^{jk_z z} \right) + \sum_{n=2}^{\infty} A_n \sin \frac{n\pi y}{a} e^{\gamma_n z}, \qquad z \leq 0 \tag{1.102}$$

Behind the inhomogeneity we have

$$E_x(y, z) = T \sin \frac{\pi y}{a} e^{-jk_z z} + \sum_{n=2}^{\infty} B_n \sin \frac{n\pi y}{a} e^{-\gamma_n z}, \qquad z \geq 0 \tag{1.103}$$

Here

$$\gamma_n = \sqrt{\frac{n^2\pi^2}{a^2} - k^2}$$

are the decay factors of the higher-order evanescent modes. To determine the unknown reflection and transmission coefficients R and T, we have to solve for the amplitudes of the higher-order modes A_n and B_n.

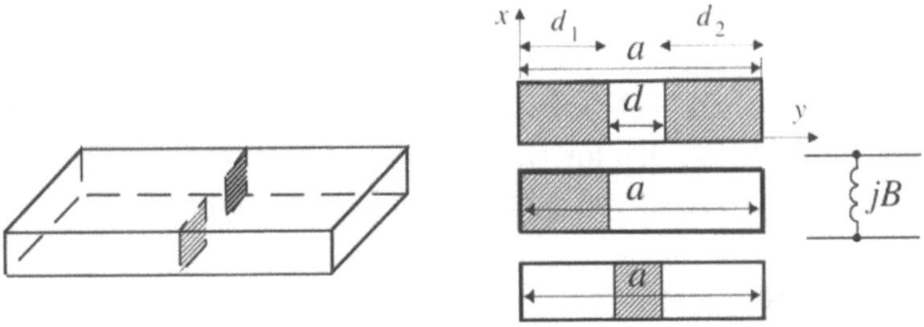

Figure 1.12. Inductive diaphragm in a rectangular waveguide and its equivalent circuit.

To this end, we should use the boundary conditions on the metal insertion. Because both expressions (1.102) and (1.103) should give the same field in the hole, we see that $T = 1 + R$ and $A_n = B_n$. We can further make use of the orthogonality property of the set of the eigenfields. Multiplying (1.102) and (1.103) at $z = 0$ by $\sin\frac{\pi y}{a}$ and integrating over y we have

$$1 + R = \frac{2}{a}\int_{d_1}^{d_1+a} E_x(y,0)\,\sin\frac{\pi y}{a}\,dy \tag{1.104}$$

Multiplying by $\sin\frac{n\pi y}{a}$ and integrating we get

$$A_n = \frac{2}{a}\int_{d_1}^{d_1+a} E_x(y,0)\,\sin\frac{n\pi y}{a}\,dy \tag{1.105}$$

In fact, $E_x(y,0)$ must be zero at the diaphragm surface. Next we should impose the continuity condition on H_y. Expressing H_y through $\mathbf{E}$ we have $H_y = j/(\omega\mu)\partial E_x/\partial z$. Using (1.102) and (1.103) we get the boundary condition

$$\sin\frac{\pi y}{a}(1-R)+\sum_{n=2}^{\infty} A_n\frac{j\gamma_n}{k_z}\sin\frac{n\pi y}{a} = \sin\frac{\pi y}{a}(1+R)-\sum_{n=2}^{\infty} A_n\frac{j\gamma_n}{k_z}\sin\frac{n\pi y}{a} \tag{1.106}$$

Coefficients A_n can be eliminated using (1.105):

$$2R\sin\frac{\pi y}{a} = 2\sum_{n=2}^{\infty}\left(\frac{2}{a}\int_0^a E_x(y,0)\,\sin\frac{n\pi y}{a}\,dy\right)\frac{j\gamma_n}{k_z}\sin\frac{n\pi y}{a} \tag{1.107}$$

This equation together with (1.104) forms a system of two integral equations, whose solution determines $E_x(y,0)$ in the hole and the reflection coefficient R. There are methods which allow analytical approximate solutions [6], or numerical methods can be used. Alternatively, multiplying (1.106) by $\sin\frac{n\pi y}{a}$ and integrating, we can arrive at an infinite system of linear equations for the unknown modal coefficients A_n which can be approximately solved using numerical methods.

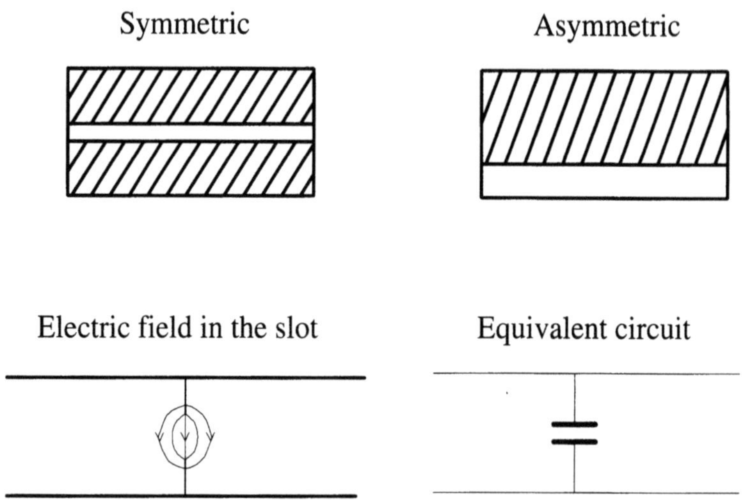

Figure 1.13. Capacitive diaphragm and its equivalent circuit.

If a diaphragm in a metal rectangular waveguide has a horizontal slot, it behaves as a shunt capacitance, Figure 1.13. In this case there is some additional electric field energy stored in the electric field in the slot.

Resonant inclusions can be obtained combining these two types of insertions, Figure 1.14. If we reduce the window area keeping the resonant frequency constant, then size b' diminishes faster than a'. In the limit we arrive at a thin horizontal slot of the length close to $\lambda_0/2$ (λ_0 is the free-space wavelength), as shown in Figure 1.14.

Connection of two waveguides with different widths is illustrated in Figure 1.15. This is equivalent to a connection of two different transmission lines with an additional inductance connected at the place of the junction.

Depending on the length, a thin conducting vertical rod ("post") in a rectangular waveguide can behave as a capacitance, inductance, or a resonant element, see Figure 1.16. For short rods $l < \lambda/4$ it is a capacitance, rods connected to the two sides of the waveguide are inductive (additional magnetic field due to its current). Long rods have

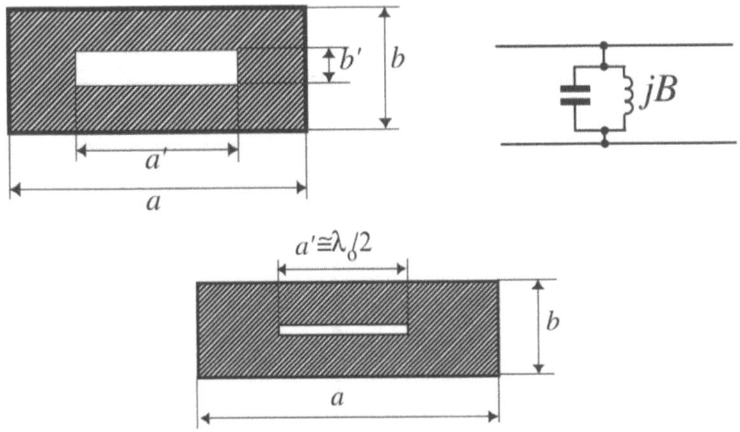

Figure 1.14. Resonant diaphragms and the corresponding equivalent circuit.

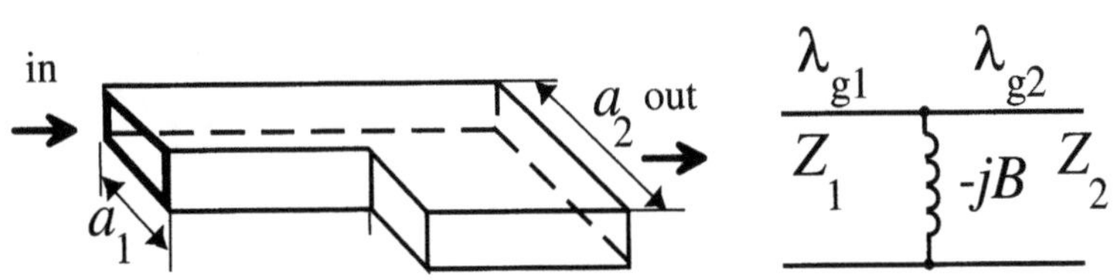

Figure 1.15. Jump in the waveguide width.

an equivalent circuit with both an inductance and a capacitance.

Matched load of a waveguide (Figure 1.17) is, obviously, equivalent to a matched transmission line.

1.6 PERIODICALLY INHOMOGENEOUS WAVEGUIDES

Waveguides with periodically repeating inhomogeneities have important applications as slow-wave structures; for instance, in amplifiers and generators. Another application is in the design of filters, because there are stop bands in the spectrum of periodical waveguides.

In periodical waveguides, the parameters are not necessarily uniform along a straight line in space (axis z) but can *periodically* depend on the z-coordinate. Of course the regular waveguide is a special case of this more general class. Let us denote the period by d, then a periodical waveguide can be defined by its complex permittivity and permeability

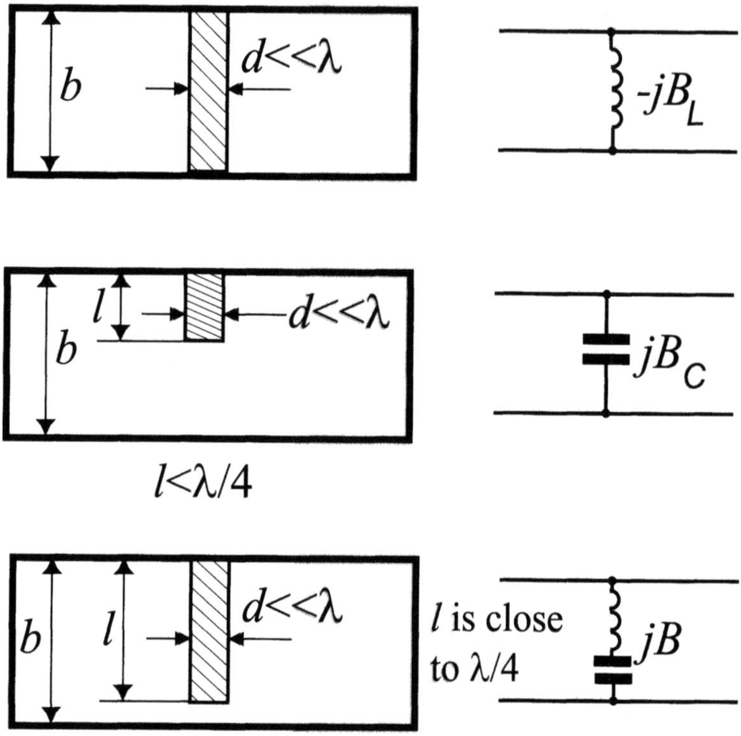

Figure 1.16. Conductive rods in a rectangular waveguide.

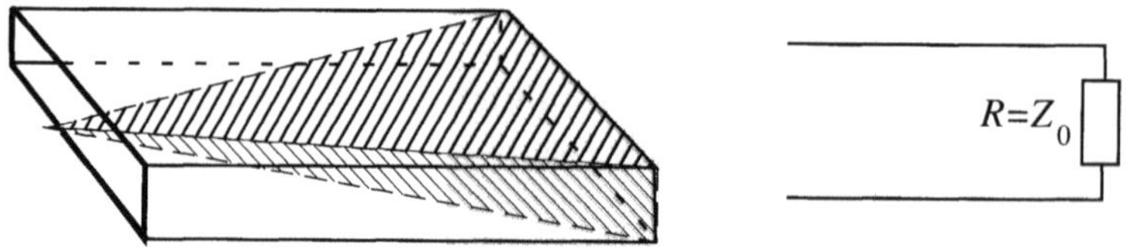

Figure 1.17. Matched load of a waveguide. The load is a pyramidal insertion made of an absorbing material.

functions which satisfy

$$\varepsilon(x, y, z + d) = \varepsilon(x, y, z), \qquad \mu(x, y, z + d) = \mu(x, y, z) \tag{1.108}$$

for all z. Waveguides with metal walls (e.g., corrugated waveguides) are also covered by this definition as the limiting case at $\varepsilon \to -j\infty$.

Waves in periodical waveguides are governed by the Maxwell equations, which are in this situation linear differential equations with *periodical coefficients.* We can make use of the Floquet theorem in the theory of differential equations to find that the solutions in periodical waveguides satisfy relations

$$\mathbf{E}(x, y, z + nd) = e^{-jnk_z d}\mathbf{E}(x, y, z)$$

$$\mathbf{H}(x, y, z + nd) = e^{-jnk_z d}\mathbf{H}(x, y, z) \tag{1.109}$$

Here $n = 0, \pm 1, \pm 2, \dots$, and k_z is a complex constant. In other words, relations (1.109) mean that the fields in the adjacent periods differ only by a constant complex multiplier. For the fields in an arbitrary point we can write

$$\mathbf{E}(x, y, z) = \mathbf{E}_0(x, y, z)e^{-jk_z z} \tag{1.110}$$

$$\mathbf{H}(x, y, z) = \mathbf{H}_0(x, y, z)e^{-jk_z z} \tag{1.111}$$

where the amplitude functions $\mathbf{E}_0$ and $\mathbf{H}_0$ are *periodical* functions with the period d. Since these functions are periodical, we can expand them into the Fourier series:

$$\mathbf{E}(x, y, z) = \sum_{m=-\infty}^{\infty} \mathbf{E}_m(x, y)\, e^{-j\left(k_z + \frac{2\pi m}{d}\right)z} \tag{1.112}$$

The members of this series are called *Floquet modes* or *spatial harmonics.*

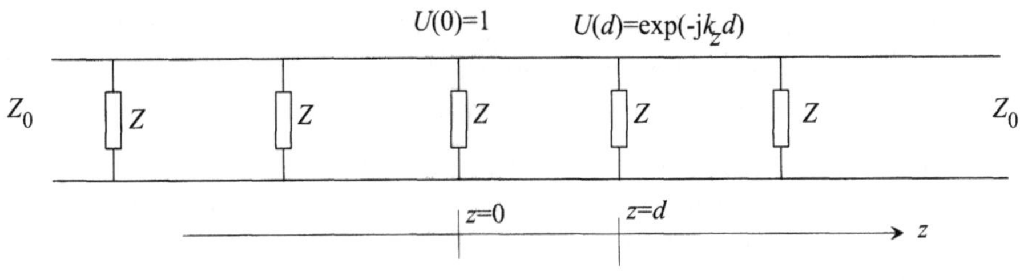

Figure 1.18. Transmission line (wave impedance Z_0) with periodical bulk insertions.

In regular waveguides functions $\mathbf{E}_0$ and $\mathbf{H}_0$ do not depend on z, and k_z is, of course, the propagation factor. A periodical variation of a waveguide leads to a "modulation" of the amplitude functions: they become periodical functions with the same period d as that of the waveguide. Every Floquet mode looks like an eigenwave in a regular waveguide: $\mathbf{E}_m(x, y)\, e^{-jk_{zm}z}$. However, taken separately one Floquet mode has little physical meaning, because only the full sum of the Fourier series (1.112) satisfies both the Maxwell equations and the boundary conditions on the waveguide walls or on interfaces.

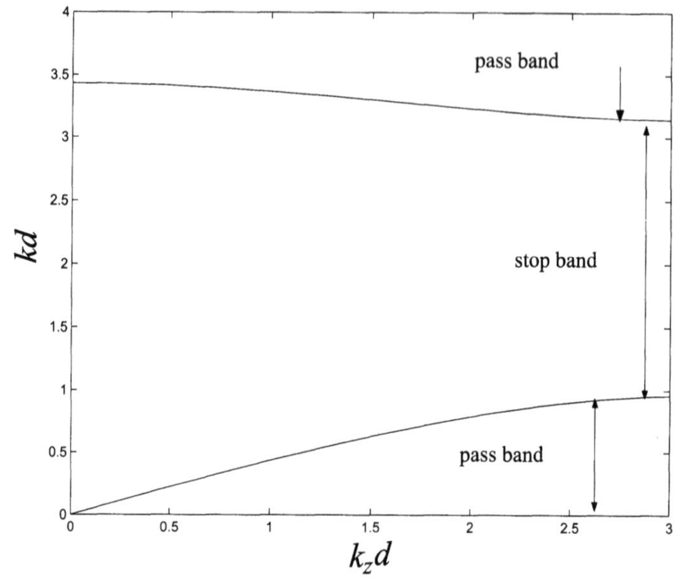

Figure 1.19. Typical dispersion curves for a periodical waveguide.

Let us consider a regular single-mode waveguide periodically perturbed by small insertions. Here, "small" means that each insertion can be represented as a bulk load in the equivalent transmission line that models the waveguide. Let us denote the line wave impedance by Z_0 and let every inclusion have the (bulk) reactive impedance Z, Figure 1.18.

The dispersion relation can be easily found in the assumption that the distance between insertions is so large that the higher-order modes decay at this distance. The derivation is based on the Floquet theorem (e.g., [8]), and leads to

$$\cos k_z d = \cos kd + j\frac{Z_0}{2Z}\sin kd \tag{1.113}$$

Dispersion properties of such waveguides are illustrated in Figure 1.19, where dispersion curves are plotted for a transmission line with capacitive loads $[Z = 1/(j\omega C)]$. At low frequencies, dispersion is weak, but at higher frequencies there is a stop band, where there are no propagating modes [the solution k_z of (1.113) is imaginary in this region]. At still higher frequencies, there are more pass bands. Varying load properties, various dispersion laws can be realized.

References

[1] D.K. Cheng, *Field and Wave Electromagnetics,* Second corrected edition, Reading, MA: Addison-Wisley, 1992.

[2] R.E. Collin, *Field Theory of Guided Waves,* Piscataway, NJ: IEEE Press, 1991.

[3] A.A. Barybin and V.A. Dmitriev, *Modern Electrodynamics and Coupled-Mode Theory: Application to Guided-Wave Optics,* Princeton, NJ: Rinton Press, 2002.

[4] L.A. Weinstein, *The Theory of Diffraction and the Factorization Method,* Boulder, CA: The Golem Press, 1969 (translation from Russian by P. Beckmann).

[5] R. Mittra and S.W. Lee, *Analytical Techniques in the Theory of Guided Waves,* New York: The Macmillan Company, 1971.

[6] L. Levin, *Theory of Waveguides: Techniques for the Solution of Waveguide Problems,* London: Newnes-Butterworth, 1975.

[7] N. Marcuvitz (ed.,) *Waveguide Handbook,* (MIT Radiation Laboratory Series, vol. 10), New York: McGraw-Hill, 1951.

[8] R.E. Collin, *Foundations for Microwave Engineering,* Piscataway, NJ: IEEE Press, 2001 (Chapter 8).

Chapter 2

Theory of high-frequency resonators

Resonators of various types are used in frequency selective devices and in material measurements. Here we present general considerations on their eigenmodes and quality factors.

2.1 MODES OF CLOSED AND OPEN RESONATORS

We start from general considerations regarding eigensolutions for the fields in various resonators. We will briefly describe how the resonant frequencies and unloaded quality factors depend on the resonator design. The fields excited in resonators by external sources can be found in terms of a combination of eigensolutions with additional gradient fields.

2.1.1 Eigensolutions

Let us consider eigensolutions for a volume closed by an ideally conducting surface and filled with a lossless isotropic material with the parameters ε and μ. The field equations are the Maxwell equations

$$\nabla \times \mathbf{E} = -\mu \frac{\partial \mathbf{H}}{\partial t}, \qquad \nabla \times \mathbf{H} = \varepsilon \frac{\partial \mathbf{E}}{\partial t} \tag{2.1}$$

with the boundary condition $\mathbf{E}_t|_s = 0$. Looking for solutions in the form

$$\mathbf{E}(x, y, z, t) = \mathbf{E}(x, y, z)e(t), \qquad \mathbf{H}(x, y, z, t) = \mathbf{H}(x, y, z)h(t) \tag{2.2}$$

and separating the variables we find that

$$\frac{1}{h(t)} \frac{de(t)}{dt} = \text{const}(t) = C_1, \qquad \frac{1}{e(t)} \frac{dh(t)}{dt} = \text{const}(t) = C_2 \tag{2.3}$$

Eliminating one of the functions we get

$$\frac{d^2e(t)}{dt^2} + \omega^2 e(t) = 0, \qquad \frac{d^2h(t)}{dt^2} + \omega^2 h(t) = 0 \tag{2.4}$$

where we denote $C_1C_2 = -\omega^2$. Thus, the general solution has the harmonic time dependence:[1]

$$e(t) = Ae^{j\omega t} + Be^{-j\omega t}, \qquad h(t) = Ce^{j\omega t} + De^{-j\omega t} \tag{2.5}$$

This actually means that we can use the complex amplitude method and write

$$\nabla \times \mathbf{E}(x,y,z) = -j\omega\mu \mathbf{H}(x,y,z), \quad \nabla \times \mathbf{H}(x,y,z) = j\omega\varepsilon \mathbf{E}(x,y,z) \tag{2.6}$$

For a closed volume with ideally conducting walls this leads to the Helmholtz equation

$$\nabla^2 \mathbf{E}(x,y,z) + k^2 \mathbf{E}(x,y,z) = 0 \tag{2.7}$$

with the boundary condition $\mathbf{n}_0 \times \mathbf{E}|_s = 0$ on the surface. We know from the mathematics that this eigenvalue problem has an infinite discrete spectrum of eigensolutions $\mathbf{E}_n$ with eigennumbers $k_n = \omega_n\sqrt{\varepsilon\mu}$. Real numbers ω_n are the resonant frequencies of the resonator.

2.1.2 Cylindrical resonators

Many resonators are actually sections of regular waveguides, and their analysis can be conveniently based on the waveguide theory.

General theory

Consider a section (length d) of an arbitrary closed cylindrical waveguide. Let the two ends have ideally conducting terminations, so that the transverse electric fields in the waveguide satisfy the boundary conditions

$$\mathbf{E}_t|_{z=0} = \mathbf{E}_t|_{z=d} = 0 \tag{2.8}$$

Assuming a single-mode regime in the waveguide we can write for the general solution in the guide

$$\mathbf{E}_t(z) = A^+\mathbf{E}_t(x,y)e^{-jk_z z} + A^-\mathbf{E}_t(x,y)e^{jk_z z} \tag{2.9}$$

[1] In the lossless system the solutions cannot decay or grow, that is why the constant C_1C_2 must be real and negative.

Applying the boundary conditions (2.8) we have

$$\mathbf{E}_t(A^+ + A^-) = 0, \qquad \text{thus} \quad A^+ = -A^- = A \tag{2.10}$$

and also

$$A\mathbf{E}_t(e^{-jk_zd} - e^{jk_zd}) = 0 \tag{2.11}$$

From here, $\sin(k_zd) = 0$, whence

$$k_z = \frac{p\pi}{d}, \qquad p = 0, 1, 2, \ldots \tag{2.12}$$

This gives the spectrum of eigensolutions, since we know from the waveguide theory that $k_z^2 = k^2 - k_c^2$, where k_c is the cut-off wavenumber for this waveguide mode. Using (2.12) we find

$$k = \sqrt{k_c^2 + k_z^2} = \sqrt{k_c^2 + \left(\frac{p\pi}{d}\right)^2}, \qquad \omega = \frac{1}{\sqrt{\varepsilon\mu}}\sqrt{k_c^2 + \left(\frac{p\pi}{d}\right)^2} \tag{2.13}$$

These frequencies are the eigenfrequencies of the resonator. Usually, they are numbered with three indices, like ω_{nmp}, because we have two indices to indicate a waveguide mode defined by its cut-off wavenumber $k_{c\{nm\}}$, and one more index p.

Rectangular and circular cross sections

For a rectangular waveguide with the cross section $a \times b$

$$k_c^2 = \left(\frac{m\pi}{a}\right)^2 + \left(\frac{n\pi}{b}\right)^2 \tag{2.14}$$

thus

$$\omega_{nmp} = \frac{1}{\sqrt{\varepsilon\mu}}\sqrt{\left(\frac{m\pi}{a}\right)^2 + \left(\frac{n\pi}{b}\right)^2 + \left(\frac{p\pi}{d}\right)^2} \tag{2.15}$$

For a circular waveguide of radius a we have

$$k_c^{\text{TM}} = \frac{p_{ns}}{a}, \qquad k_c^{\text{TE}} = \frac{p'_{ns}}{a} \tag{2.16}$$

where p_{ns} and p'_{ns} are the roots of the equations $J_n(k_ca) = 0$ and $J'_n(k_ca) = 0$, respectively. Here, J_n is the Bessel function. Thus, for resonators with the circular cross section

$$\omega_{nsp} = \frac{1}{\sqrt{\varepsilon\mu}}\sqrt{\left(\frac{p_{ns}}{a}\right)^2 + \left(\frac{p\pi}{d}\right)^2}, \quad \text{TM modes} \tag{2.17}$$

$$\omega_{nsp} = \frac{1}{\sqrt{\varepsilon\mu}}\sqrt{\left(\frac{p'_{ns}}{a}\right)^2 + \left(\frac{p\pi}{d}\right)^2}, \quad \text{TE modes} \tag{2.18}$$

2.1.3 Mode orthogonality

Let us denote by $\mathbf{E}_i$ and $\mathbf{E}_j$ the electric fields of two different resonator modes (the corresponding eigenfrequencies $\omega_i \neq \omega_j$) in a closed resonator. Our next goal is to prove that they are orthogonal, i.e.,

$$\int_V \varepsilon \mathbf{E}_i \cdot \mathbf{E}_j \, dV = 0 \tag{2.19}$$

where the integral is taken over the volume of the resonator. The idea of the proof is to reduce the volume integral to surface integrals and make use of the boundary conditions.

The fields of the two modes satisfy the Maxwell equations

$$\nabla \times \mathbf{E}_i = -j\omega_i \mu \mathbf{H}_i, \qquad \nabla \times \mathbf{E}_j = -j\omega_j \mu \mathbf{H}_j \tag{2.20}$$

We scalarly multiply the first equation by $\mathbf{H}_j$ and the second one by $\mathbf{H}_i$:

$$\mathbf{H}_j \cdot \nabla \times \mathbf{E}_i = -j\omega_i \mu \mathbf{H}_i \cdot \mathbf{H}_j, \qquad \mathbf{H}_i \cdot \nabla \times \mathbf{E}_j = -j\omega_j \mu \mathbf{H}_i \cdot \mathbf{H}_j \tag{2.21}$$

Next, we make use of the identity

$$\nabla \cdot (\mathbf{E}_i \times \mathbf{H}_j) = \mathbf{H}_j \cdot \nabla \times \mathbf{E}_i - \mathbf{E}_i \cdot \nabla \times \mathbf{H}_j \tag{2.22}$$

(and a similar one for $\mathbf{E}_j$ and $\mathbf{H}_i$) and rewrite our result as

$$\nabla \cdot (\mathbf{E}_i \times \mathbf{H}_j) + \mathbf{E}_i \cdot \nabla \times \mathbf{H}_j = -j\omega_i \mu \mathbf{H}_i \cdot \mathbf{H}_j \tag{2.23}$$

$$\nabla \cdot (\mathbf{E}_j \times \mathbf{H}_i) + \mathbf{E}_j \cdot \nabla \times \mathbf{H}_i = -j\omega_j \mu \mathbf{H}_i \cdot \mathbf{H}_j \tag{2.24}$$

Because $\nabla \times \mathbf{H}_{i,j} = j\omega_{i,j} \varepsilon \mathbf{E}_{i,j}$, we get

$$\nabla \cdot (\mathbf{E}_i \times \mathbf{H}_j) + j\omega_j \varepsilon \mathbf{E}_i \cdot \mathbf{E}_j = -j\omega_i \mu \mathbf{H}_i \cdot \mathbf{H}_j \tag{2.25}$$

$$\nabla \cdot (\mathbf{E}_j \times \mathbf{H}_i) + j\omega_i \varepsilon \mathbf{E}_i \cdot \mathbf{E}_j = -j\omega_j \mu \mathbf{H}_i \cdot \mathbf{H}_j \tag{2.26}$$

Next we of course integrate these relations over the resonator volume. Applying the Gauss theorem to the divergence terms we see that their contribution vanishes because of the boundary condition.[2] Thus, we have a system of two linear equations

$$\omega_j \int_V \varepsilon \mathbf{E}_i \cdot \mathbf{E}_j \, dV = -\omega_i \int_V \mu \mathbf{H}_i \cdot \mathbf{H}_j \, dV \tag{2.27}$$

[2]This works for nonideally conducting (lossy) walls as well: it is enough that the fields vanish somewhere inside the walls. It is only important that the resonator is *closed*.

$$\omega_i \int_V \varepsilon \mathbf{E}_i \cdot \mathbf{E}_j \, dV = -\omega_j \int_V \mu \mathbf{H}_i \cdot \mathbf{H}_j \, dV \tag{2.28}$$

Its determinant is $\omega_j^2 - \omega_i^2 \neq 0$, whence the system has only a trivial solution:

$$\int_V \varepsilon \mathbf{E}_i \cdot \mathbf{E}_j \, dV = 0, \qquad \int_V \mu \mathbf{H}_i \cdot \mathbf{H}_j \, dV = 0 \tag{2.29}$$

Note that if we repeat the same derivation for just one mode, i.e., assuming $\omega_i = \omega_j = \omega$ and $\mathbf{E}_i = \mathbf{E}_j = \mathbf{E}$, $\mathbf{H}_i = \mathbf{H}_j = \mathbf{H}$, then it follows that

$$\int_V \varepsilon \mathbf{E} \cdot \mathbf{E} \, dV = -\int_V \mu \mathbf{H} \cdot \mathbf{H} \, dV \tag{2.30}$$

2.1.4 Losses in resonators. Quality factor

Because due to losses fields in a resonator exponentially decay, if not supported by a source, the losses in a resonator can be described by introducing complex resonant frequencies (eigenfrequencies)[3] $\omega = \omega' + j\omega''$. The quality factor is defined as in the circuit theory:

$$Q = \frac{\omega' W}{P} \tag{2.31}$$

where W is the field energy stored in the resonator, and P is the loss power (energy lost during one second). The field energy is the volume integral of the absolute value of the electric field *squared*, thus, it depends on time as $W \sim \exp(-2\omega'' t)$. From here,

$$P = -\frac{dW}{dt} = 2\omega'' W \tag{2.32}$$

Thus,

$$Q = \frac{\omega' W}{P} = \frac{\omega'}{2\omega''} \tag{2.33}$$

and

$$\omega = \omega' \left(1 + j\frac{1}{2Q}\right) \tag{2.34}$$

In closed waveguides, losses in metal walls usually dominate over the losses in the resonator filling. Consider the loss factor due to losses in metal walls in more detail:

$$P = \frac{R_s}{2} \oint_S |J_s|^2 \, dS = \frac{R_s}{2} \oint_S |H_t|^2 \, dS \tag{2.35}$$

[3] Of course the signal frequency remains real.

Here $R_s = \sqrt{\omega\mu_{\text{metal}}/2\sigma}$ is the surface resistance, and $\mathbf{H}_t$ is the tangential to the wall magnetic field component. To find the quality factor we have to evaluate the stored energy W:

$$W = W_m + W_e = \frac{\mu}{4}\int_V |H|^2\, dV + \frac{\varepsilon}{4}\int_V |E|^2\, dV \tag{2.36}$$

These two components are equal [see the previous section, equation (2.30)], thus

$$W = \frac{\mu}{2}\int_V |H|^2\, dV \tag{2.37}$$

Finally,

$$Q = \frac{\omega'\mu}{R_s}\frac{\int_V |H|^2\, dV}{\oint_S |H_t|^2\, dS} \tag{2.38}$$

Let us rewrite this result in terms of the skin depth

$$\delta = \sqrt{\frac{2}{\omega\sigma\mu_{\text{metal}}}} \tag{2.39}$$

If $\mu = \mu_{\text{metal}}$, which is usually the case,

$$\frac{\omega'\mu}{R_s} = \sqrt{2\omega\sigma\mu_{\text{metal}}} = \frac{2}{\delta} \tag{2.40}$$

Substituting in (2.38), we get

$$Q = \frac{\int_V |H|^2\, dV}{\frac{\delta}{2}\oint_S |H_t|^2\, dS} \tag{2.41}$$

2.2 EXCITATION OF RESONATORS

Simple and general expressions exist for calculation of the amplitude of the field in a resonator excited by given sources (external currents in a probe or external fields at an opening in the wall). This theory is based on the modal expansion of the waveguide fields.

2.2.1 Eigenfunction expansion

Consider a *closed* resonator. The set of its eigenfunctions $\mathbf{E}_n$ forms a closed system of orthogonal functions, thus we can expand the field excited by a source current $\mathbf{J}$ as

$$\mathbf{E} = \sum A_n \mathbf{E}_n \tag{2.42}$$

where A_n are complex scalar amplitude coefficients. The field in the resonator satisfies

$$\nabla \times \left(\frac{1}{\mu}\nabla \times \mathbf{E}\right) - \omega^2 \varepsilon \mathbf{E} = -j\omega \mathbf{J} \tag{2.43}$$

The eigenfunctions satisfy the homogeneous equation

$$\nabla \times \left(\frac{1}{\mu}\nabla \times \mathbf{E}_n\right) - \omega_n^2 \varepsilon \mathbf{E}_n = 0 \tag{2.44}$$

Substitute expansion (2.42) into the field equation (2.43). The result is[4]

$$\sum A_n \left[\nabla \times \left(\frac{1}{\mu}\nabla \times \mathbf{E}_n\right) - \omega^2 \varepsilon \mathbf{E}_n\right] = -j\omega \mathbf{J} \tag{2.45}$$

From (2.44) we see that

$$\nabla \times \left(\frac{1}{\mu}\nabla \times \mathbf{E}_n\right) = \omega_n^2 \varepsilon \mathbf{E}_n \tag{2.46}$$

thus, (2.45) is equivalent to

$$\sum A_n(\omega_n^2 - \omega^2)\varepsilon \mathbf{E}_n = -j\omega \mathbf{J} \tag{2.47}$$

Multiplication by $\mathbf{E}_m$ and integration over the resonator volume gives finally

$$A_m = \frac{-j\omega}{\omega_m^2 - \omega^2} \frac{\int_V \mathbf{J} \cdot \mathbf{E}_m \, dV}{\int_V \varepsilon \mathbf{E}_m \cdot \mathbf{E}_m \, dV} \tag{2.48}$$

Here we have made use of the orthogonality property (2.19) of the eigenfields of different resonator modes.

Strictly speaking, the above theory is correct for closed resonators only. However, any resonator is always connected to the outside world, at least through feeding lines. Consider where is the fault. Taking divergence of (2.42), we see that our solution is divergence free: $\nabla \cdot \mathbf{E} = 0$, because each term in the series satisfies the homogeneous Maxwell equations. Thus, the solution cannot be correct if $\nabla \cdot \mathbf{J} \neq 0$. It appears at first sight that we have no such problem with the magnetic field, because $\nabla \cdot \mathbf{B} = 0$, anyway. However, it is not so in our case, because

[4]This assumes that the material parameters of the waveguide filling do not depend on the frequency within the resonance band, which is a limitation of this method. This is because in equation (2.44) the material parameters should be taken at the resonant frequency ω_n, but the fields oscillate at the signal frequency ω.

a resonator can have holes in the walls, and magnetic field lines can go out from the resonator volume or come in, thus creating a source for the magnetic field inside the resonator volume. In other words, the divergence of the magnetic field does not have to be zero at resonator openings.

This defect can be corrected by adding gradient fields to expansion (2.42), so that the solution is thought as a sum

$$\mathbf{E}_{\text{full}} = \mathbf{E} + \mathbf{E}_g = \sum A_n \mathbf{E}_n + \mathbf{E}_g \tag{2.49}$$

where $\mathbf{E}$ satisfies $\nabla \cdot \mathbf{E} = 0$ and is given by (2.42) and (2.48), and $\mathbf{E}_g$ satisfies $\nabla \times \mathbf{E}_g = 0$. The solution for the gradient field can be expressed in terms of a scalar potential: $\mathbf{E}_g = -\nabla\phi$, and the equation for ϕ is obviously the static equation.

For many applications this correction is not important, since the gradient field has no resonances. However, it becomes very important if we calculate the field on an aperture, for example to find the input impedance of a resonator. The field distribution near apertures and probes is, of course, affected by the gradient field component $\mathbf{E}_g$.

2.2.2 Excitation of resonators as sections of waveguides

In most cases a resonator can be considered as a section of a waveguide (not necessarily closed by boundaries: The terminations may be partially transparent). In this case, the theory of resonator excitation can be built using the theory of waveguide excitation [1]. Suppose that the resonator volume is a waveguide section from $z = 0$ to $z = L$. Let us denote by $\mathbf{E}_n$, $\mathbf{H}_n$ the fields of the eigenmode number n propagating in the positive direction of z, and by $\mathbf{E}_{-n}$, $\mathbf{H}_{-n}$ the fields of the same mode traveling in the opposite direction. The frequency of these modes is equal to the excitation frequency ω. The corresponding propagation constants are $k_{zn} = -k_{z(-n)}$. Next, we denote

$$\mathbf{E}^r_n = \mathbf{E}_n + R_n \mathbf{E}_{-n}, \qquad \mathbf{H}^r_n = \mathbf{H}_n + R_n \mathbf{H}_{-n} \tag{2.50}$$

where R_n is the reflection coefficient for mode n at $z = L$. Also,

$$\mathbf{E}^r_{-n} = \mathbf{E}_{-n} + R_{-n} \mathbf{E}_n, \qquad \mathbf{H}^r_{-n} = \mathbf{H}_{-n} + R_{-n} \mathbf{H}_n \tag{2.51}$$

where R_{-n} is the reflection coefficient for mode n at $z = 0$.

Fields (2.50) and (2.51) satisfy the Maxwell equations at the excitation frequency[5] and the boundary conditions on the side walls. Suppose that the sources are located in the region $z_1 < z < z_2$, where $z_1 > 0$ and $z_2 < L$. In this situation it is natural to construct the solution for the field excited by these sources in the resonator as

$$\mathbf{E} = \sum_n C_n \mathbf{E}^r_n, \qquad \mathbf{H} = \sum_n C_n \mathbf{H}^r_n \tag{2.52}$$

for $z_2 < z < L$ and as

$$\mathbf{E} = \sum_n C_{-n} \mathbf{E}^r_{-n}, \qquad \mathbf{H} = \sum_n C_{-n} \mathbf{H}^r_{-n} \tag{2.53}$$

for $0 < z < z_1$. Here constants C_n and C_{-n} are the amplitude coefficients which we have to find.

This is easy to do using the orthogonality relation

$$\int_S (\mathbf{E}^r_n \times \mathbf{H}^r_m - \mathbf{E}^r_m \times \mathbf{H}^r_n) \cdot d\mathbf{S} = 0 \qquad \text{for} \quad m \neq n \tag{2.54}$$

which follows from the orthogonality of the modal fields of closed waveguides. In the same way as in the theory of waveguide excitation, the following result can be obtained:

$$C_n = \frac{1}{N^r_n} \int_V \mathbf{J} \cdot \mathbf{E}^r_{-n}\, dV, \qquad C_{-n} = \frac{1}{N^r_n} \int_V \mathbf{J} \cdot \mathbf{E}^r_n\, dV \tag{2.55}$$

where $\mathbf{J}$ is the given source current density and the integration is taken over the volume $z_1 < z < z_2$, where $\mathbf{J} \neq 0$. The norm N^r_n reads

$$N^r_n = \int_S (\mathbf{E}^r_n \times \mathbf{H}^r_{-n} - \mathbf{E}^r_{-n} \times \mathbf{H}^r_n) \cdot d\mathbf{S} \tag{2.56}$$

After substitution of (2.50) and (2.51) this reduces to

$$N^r_n = (1 - R_n R_{-n}) N_n \tag{2.57}$$

where N_n is the norm of the waveguide eigenmode n:

$$N_n = \int_S (\mathbf{E}_n \times \mathbf{H}_{-n} - \mathbf{E}_{-n} \times \mathbf{H}_n) \cdot d\mathbf{S} \tag{2.58}$$

According to (2.55), the resonant phenomena take place when the norm N^r_n becomes small. Consider, for example, a resonator with ideally conducting walls at $z = 0$ and $z = L$. In this case $R_{-n} = -1$ and $R_n = -e^{2jk_{zn}L}$. Thus, $1 - R_n R_{-n} = -2je^{jk_{zn}L}\sin(k_{zn}L)$. Obviously, we will observe resonances at $k_{zn}L = m\pi$, $m = 0, 1, 2, \ldots$

[5]Note that in this method there is no restriction on the resonator filling, as it was in the previous section (dispersionless filling).

References

[1] L.A. Vainstein, *Electromagnetic Waves,* Second edition, Moscow: Radio i Sviaz, 1988 (in Russian).

Chapter 3

Waves in crystals and anisotropic waveguides

Many materials used in millimeter-wave techniques including fabrication of waveguides are anisotropic crystals. We start our study of waves in anisotropic media and anisotropic waveguides from a discussion of the electromagnetic properties of crystals.

3.1 ELECTROMAGNETIC PROPERTIES OF ANISOTROPIC CRYSTALS. RECIPROCITY

The most important feature affecting electromagnetic waves in most crystals is their *anisotropy.* This means that electric fields applied in different directions cause different polarizations of the medium. In electromagnetic description of material properties, this means that the permittivity and permeability are in general *tensors* or *dyadics* which define general linear transformation of vectors:

$$\mathbf{D} = \overline{\overline{\varepsilon}}(\omega) \cdot \mathbf{E}, \qquad \mathbf{B} = \overline{\overline{\mu}}(\omega) \cdot \mathbf{H} \tag{3.1}$$

Three-dimensional algebraic linear operators $\overline{\overline{\varepsilon}}$ and $\overline{\overline{\mu}}$ can be considered as

- *Tensors,* and written in index notation as ε_{ij}, μ_{ij} (indices i, j take values $1, 2, 3$)
- *Matrices,* and expressed in a Cartesian coordinate system as

$$\begin{pmatrix} \varepsilon_{xx} & \varepsilon_{xy} & \varepsilon_{xz} \\ \varepsilon_{yx} & \varepsilon_{yy} & \varepsilon_{yz} \\ \varepsilon_{zx} & \varepsilon_{zy} & \varepsilon_{zz} \end{pmatrix}$$

- *Dyadics,*[1] written as linear combinations of pairs of vectors:

$$\overline{\overline{\varepsilon}} = \sum_i \mathbf{a}_i \mathbf{b}_i, \qquad \overline{\overline{\mu}} = \sum_i \mathbf{c}_i \mathbf{d}_i$$

The time-reversal symmetry of the Maxwell equations imposes some restrictions on the material property tensors. Consider the microscopic Maxwell equation (for elementary particles in vacuum)

$$\nabla \times \mathbf{E} = -\frac{\partial}{\partial t}\mu_0 \mathbf{H} \tag{3.2}$$

Let us *postulate* that the electric charge does not change if one (formally) reverses time, and let us *reverse the direction of time.* We observe that as the charge does not change, the electric field does not change, although the time flows back now. Next, operator nabla ∇ remains the same, since it contains differentiations with respect to spatial coordinates only. Thus, the left-hand side of (3.2) is invariant under the time-reversal operation. On the right-hand side of that equation time t changes sign, of course. But also $\mathbf{H}$ changes sign: magnetic field is created by *moving* charges, and they now move in the opposite direction, as the time has been reversed. Thus, equation (3.2) does not change if the time direction is reversed. The same observations can be done regarding the second Maxwell equation. We conclude that the microscopic electromagnetic processes are invariant with respect to the time reversal.

All materials consist of elementary particles, and if we are interested in linear phenomena, these particles move as dictated by the microscopic Maxwell equations. If there are no other forces besides the electromagnetic interactions between these particles as described by the Maxwell equations, this symmetry property is also present at the macroscopic level.

On the level of radio engineering and applications, this is reflected in the *reciprocity* property of materials and devices. If the system is symmetric with respect to the time inverse, the reciprocity theorem is valid.[2] We conclude that if there is no external force which breaks the natural symmetry of the Maxwell equations, materials must be reciprocal. From the reciprocity theorem (Appendix B) we know that in reciprocal

---

[1] Basic definitions and properties of dyadics can be found in Appendix A.

[2] The proof of this statement is not trivial and can be found in the physics literature [1–3].

media material parameters are *symmetric matrices.* Thus, in crystals we normally have

$$\overline{\overline{\varepsilon}} = \overline{\overline{\varepsilon}}^T \tag{3.3}$$

where T denotes transposition. In other words, the permittivity matrix has the form

$$\begin{pmatrix} \varepsilon_{xx} & \varepsilon_{xy} & \varepsilon_{xz} \\ \varepsilon_{xy} & \varepsilon_{yy} & \varepsilon_{yz} \\ \varepsilon_{xz} & \varepsilon_{yz} & \varepsilon_{zz} \end{pmatrix} \tag{3.4}$$

The same is true for the permeability matrix.

Exceptions from this rule happen when the material parameters $\overline{\overline{\varepsilon}}$ and (or) $\overline{\overline{\mu}}$ depend on an *external* force which is time-odd. Most often such force is an external magnetic field (remember that magnetic fields change sign under time reversal). *Magnetized* plasmas and ferrites are nonreciprocal. Also, there are so called magnetic crystals with spontaneous magnetic properties. They are nonreciprocal as well.

Next, we will study electromagnetic waves in reciprocal (nonmagnetic) crystals. Nonreciprocal media, in particular magnetized ferrites, are considered in Chapter 4.

3.2 ELECTROMAGNETIC WAVES IN NONMAGNETIC CRYSTALS

In reciprocal crystals, the material tensors (matrices, dyadics) are symmetric. From mathematics we know that for any real symmetric matrix there exists a coordinate system in which this matrix has the diagonal form

$$\begin{pmatrix} \varepsilon_1 & 0 & 0 \\ 0 & \varepsilon_2 & 0 \\ 0 & 0 & \varepsilon_3 \end{pmatrix} \tag{3.5}$$

The axes of the coordinate system in which the permittivity tensor is diagonal (here we assume that $\overline{\overline{\mu}}$ is a scalar) are called the *main axes.*

Definition. *If $\varepsilon_1 = \varepsilon_2 = \varepsilon_3 = \varepsilon$, the material is called isotropic. In this case, actually, the tensor is reduced to a scalar. No preferred direction exists in space.*

Definition. *If $\varepsilon_1 = \varepsilon_2 \neq \varepsilon_3$, the material is called uniaxial. One preferred direction exists, that along axis 3. In planes orthogonal to that axis the properties of the material are isotropic.*

Definition. *If $\varepsilon_1 \neq \varepsilon_2 \neq \varepsilon_3$, the material is called biaxial. In the plane orthogonal to axis 3, there is another preferred direction defined* *by axis 1 or 2.* This is the most general case if the medium is reciprocal and lossless. Thus, to describe electromagnetic properties of anisotropic dielectrics we need at most three scalar parameters. If the permeability of a reciprocal medium is also a tensor, that can be written in the diagonal form as well. The problem here is that the coordinate system in which the permeability is diagonal does not always coincide with that for the permittivity.

This sounds quite general and simple, but we should remember that the constitutive dyadics for lossy (dispersive) media are complex matrices. Actually, we can write them as sums of two real matrices (for the real and imaginary parts). Both are symmetric, but only if both the real and imaginary parts have the same eigenvectors, that is, if they take diagonal forms in the same coordinate system, the notion of uniaxial and biaxial media has a physical sense. Otherwise, reducing only one of the parts to the diagonal form does not really simplify the analysis. However, in the optics of transparent crystals and for low-loss electromagnetic materials, where the imaginary parts can be neglected, this classification is extremely useful.

3.2.1 Plane waves. Poynting vector

Consider plane waves in crystals. Assuming the plane-wave *Ansatz* $e^{-j\mathbf{k}\cdot\mathbf{r}}$ we have, from the Maxwell equations,

$$\mathbf{k}\times\mathbf{E} = \omega\mathbf{B}, \quad \mathbf{k}\times\mathbf{H} = -\omega\mathbf{D}, \qquad \mathbf{k}\cdot\mathbf{D} = 0, \quad \mathbf{k}\cdot\mathbf{B} = 0 \tag{3.6}$$

For homogeneous plane waves, that is, for real vectors $\mathbf{k}$, we observe that the phase propagation direction (defined by vector $\mathbf{k}$) is orthogonal to vectors $\mathbf{D}$ and $\mathbf{B}$. On the other hand, $\mathbf{E}$ is orthogonal to $\mathbf{B}$, and $\mathbf{D}$ is orthogonal to $\mathbf{H}$.

Let us now find the energy flow direction. Consider the standard derivation of the Poynting theorem: combining the two curl Maxwell equations, we have, in source-free regions,

$$\mathbf{E}\cdot\nabla\times\mathbf{H} - \mathbf{H}\cdot\nabla\times\mathbf{E} = \mathbf{E}\cdot\frac{\partial\mathbf{D}}{\partial t} + \mathbf{H}\cdot\frac{\partial\mathbf{B}}{\partial t} \tag{3.7}$$

This can be interpreted as the power conservation relation, because the left-hand side is equal to the divergence of a vector:

$$\mathbf{E}\cdot\nabla\times\mathbf{H} - \mathbf{H}\cdot\nabla\times\mathbf{E} = -\nabla\cdot(\mathbf{E}\times\mathbf{H}) = -\nabla\cdot\mathbf{S} \tag{3.8}$$

(which is the same as in vacuum or in an isotropic medium). Now, we should express the right-hand side of (3.7) as the time derivative of a scalar. Then, that scalar will have the meaning of the power density. Indeed, the left-hand side tells how much of something (at this stage we just guess that this is *power*) flows out or into an element of volume, then the right-hand side should be equal to the increase or decrease of the energy stored in this volume.

For reciprocal crystals, this can be done easily. Consider the time derivative of $\mathbf{E} \cdot \mathbf{D}$, assuming material relations (3.1):

$$\frac{\partial}{\partial t}(\mathbf{E} \cdot \mathbf{D}) = \frac{\partial \mathbf{E}}{\partial t} \cdot (\overline{\overline{\varepsilon}} \cdot \mathbf{E}) + \mathbf{E} \cdot \frac{\partial(\overline{\overline{\varepsilon}} \cdot \mathbf{E})}{\partial t} = 2\mathbf{E} \cdot \frac{\partial \mathbf{D}}{\partial t} \tag{3.9}$$

Here, we have essentially used that in reciprocal crystals the permittivity tensor is symmetric: $\overline{\overline{\varepsilon}} = \overline{\overline{\varepsilon}}^T$, and that the medium is stationary.[3] Otherwise, the previous result would be wrong. Doing the same for vectors $\mathbf{H}$ and $\mathbf{B}$, we conclude that

$$\mathbf{E} \cdot \frac{\partial \mathbf{D}}{\partial t} + \mathbf{H} \cdot \frac{\partial \mathbf{B}}{\partial t} = \frac{1}{2}\frac{\partial}{\partial t}(\mathbf{E} \cdot \mathbf{D} + \mathbf{H} \cdot \mathbf{B}) \tag{3.10}$$

Thus, the Poynting vector is

$$\mathbf{S} = \mathbf{E} \times \mathbf{H} \tag{3.11}$$

and the energy density reads

$$W = \frac{1}{2}(\mathbf{E} \cdot \mathbf{D} + \mathbf{H} \cdot \mathbf{B}) \tag{3.12}$$

Here we have been working in the time domain. For the frequency domain the complex Poynting vector is introduced in the usual way. Let us stress once more that in more complex media these formulas for the Poynting vector and the energy density are not correct.

Finally, we see that if a plane wave propagates in an anisotropic medium, the direction of the power flow is different from the direction of the wave vector. This is illustrated in Figure 3.1.

3.2.2 Eigenwaves in uniaxial crystals

Considering uniaxial crystals, let us denote $\varepsilon_1 = \varepsilon_2 = \varepsilon_t$ (the transverse permittivity) and direct the axis z of a Cartesian coordinate system

---

[3]Meaning that the permittivity $\overline{\overline{\varepsilon}}$ does not depend on time.

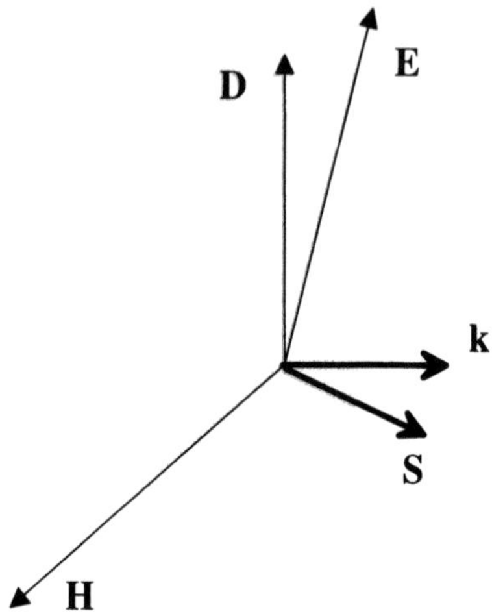

Figure 3.1. Field vectors of a homogeneous plane wave in an anisotropic dielectric ($\mathbf{B} = \mu_0 \mathbf{H}$). The Poynting vector $\mathbf{S} = \mathbf{E} \times \mathbf{H}$ is orthogonal to $\mathbf{E}$ and $\mathbf{H}$, while the wave vector $\mathbf{k}$ is orthogonal to $\mathbf{H}$ and $\mathbf{D}$.

along the main axis 3. $\overline{\overline{\varepsilon}}$ and $\overline{\overline{\mu}}$ are uniaxial dyadics with transverse (t) and longitudinal components (z):

$$\overline{\overline{\varepsilon}} = \varepsilon_0(\varepsilon_t \overline{\overline{I}}_t + \varepsilon_z \mathbf{z}_0\mathbf{z}_0), \qquad \overline{\overline{\mu}} = \mu_0(\mu_t \overline{\overline{I}}_t + \mu_z \mathbf{z}_0\mathbf{z}_0) \tag{3.13}$$

Here $\mathbf{z}_0$ stands for the unit vector along the crystal axis (z-axis). $\overline{\overline{I}}_t = \mathbf{x}_0\mathbf{x}_0 + \mathbf{y}_0\mathbf{y}_0$ is the transverse unit dyadic.

To study plane electromagnetic waves in uniaxial media we Fourier transform the Maxwell equations in the plane orthogonal to the axis. This means that we look for solutions which depend on the coordinates as $e^{-j(k_x x + k_y y)} f(z)$ or, more precisely, we expand the solution into a two-dimensional Fourier integral. The two-dimensional Fourier variable we denote by $\mathbf{k}_t = k_x \mathbf{x}_0 + k_y \mathbf{y}_0$. The nabla operator is then replaced by

$$\nabla \rightarrow -j\mathbf{k}_t + \frac{\partial}{\partial z}\mathbf{z}_0 \tag{3.14}$$

Splitting the fields into the normal and transverse parts (our system is a *waveguide* along z) like in Section 1.1,

$$\mathbf{E} = \mathbf{E}_t + E_z \mathbf{z}_0, \qquad \mathbf{H} = \mathbf{H}_t + H_z \mathbf{z}_0 \tag{3.15}$$

the Maxwell equations take the form

$$-j\mathbf{k}_t \times \mathbf{E} + \frac{\partial}{\partial z}\mathbf{z}_0 \times \mathbf{E}_t = -j\omega \overline{\overline{\mu}} \cdot \mathbf{H}$$

$$-j\mathbf{k}_t \times \mathbf{H} + \frac{\partial}{\partial z}\mathbf{z}_0 \times \mathbf{H}_t = j\omega\overline{\overline{\varepsilon}} \cdot \mathbf{E} \tag{3.16}$$

Similarly to isotropic waveguides, the longitudinal field components can be expressed in terms of the transverse fields:

$$E_z\mathbf{z}_0 = -\frac{1}{\omega\varepsilon_0\varepsilon_z}\mathbf{k}_t \times \mathbf{H}_t, \qquad H_z\mathbf{z}_0 = \frac{1}{\omega\mu_0\mu_z}\mathbf{k}_t \times \mathbf{E}_t \tag{3.17}$$

and eliminated, which converts (3.16) into the system of two vector transmission-line equations

$$\frac{\partial}{\partial z}\mathbf{E}_t = \left(j\omega\mu_0\mu_t\overline{\overline{I}}_t - \frac{j}{\omega\varepsilon_0\varepsilon_z}\mathbf{k}_t\mathbf{k}_t\right) \cdot \mathbf{z}_0 \times \mathbf{H}_t$$

$$\frac{\partial}{\partial z}\mathbf{z}_0 \times \mathbf{H}_t = \left(j\omega\varepsilon_0\varepsilon_t\overline{\overline{I}}_t - \frac{j}{\omega\mu_0\mu_z}\mathbf{z}_0 \times \mathbf{k}_t\,\mathbf{z}_0 \times \mathbf{k}_t\right) \cdot \mathbf{E}_t \tag{3.18}$$

The second-order wave equation for the transverse electric field component $\mathbf{E}_t$ immediately follows from the transmission-line equations (3.18) after elimination of $\mathbf{z}_0 \times \mathbf{H}_t$, and it takes the form

$$\frac{\partial^2}{\partial z^2}\mathbf{E}_t + \left(\beta_{TM}^2\frac{\mathbf{k}_t\mathbf{k}_t}{k_t^2} + \beta_{TE}^2\frac{\mathbf{z}_0 \times \mathbf{k}_t\;\mathbf{z}_0 \times \mathbf{k}_t}{k_t^2}\right) \cdot \mathbf{E}_t = 0 \tag{3.19}$$

Because the dyadic in the last equation is diagonal, the eigensolutions of (3.19) are two linearly polarized vectors: one is proportional to $\mathbf{k}_t$ and the other one — to $\mathbf{z}_0 \times \mathbf{k}_t$. Indeed, if the tangential electric field vector $\mathbf{E}_t$ is parallel to $\mathbf{k}_t$, the wave equation becomes

$$\frac{\partial^2}{\partial z^2}E_t + \beta_{TM}^2 E_t = 0 \tag{3.20}$$

If it is parallel to $\mathbf{z}_0 \times \mathbf{k}_t$, we have

$$\frac{\partial^2}{\partial z^2}E_t + \beta_{TE}^2 E_t = 0 \tag{3.21}$$

The first solution corresponds to a TM-polarized wave, with the magnetic field orthogonal to the vector $\mathbf{k}_t$, and the second one is a TE-wave. β_{TM} and β_{TE} are the longitudinal components of the propagation factors for the TM- and TE-polarized eigenwaves $e^{-j\beta_{TM,TE}z}$, respectively:

$$\beta_{TM}^2 = \frac{\varepsilon_t}{\varepsilon_z}\left(k_0^2\varepsilon_z\mu_t - k_t^2\right) \tag{3.22}$$

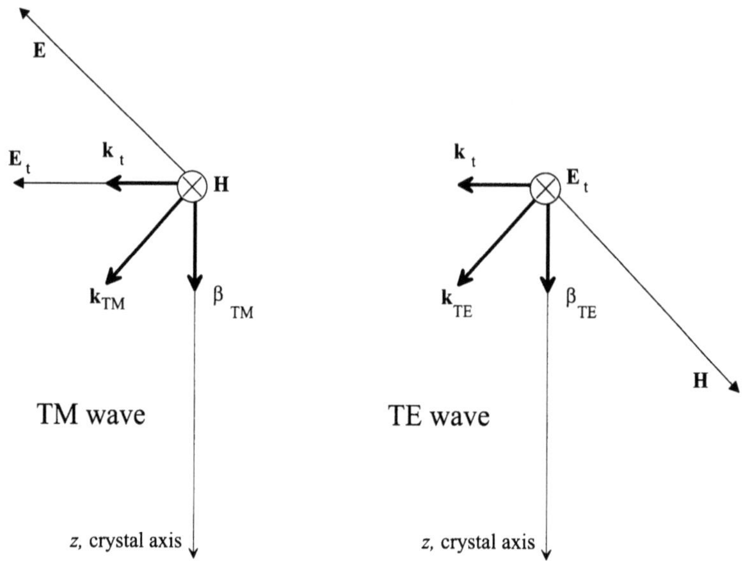

Figure 3.2. TM and TE modes in uniaxial crystals.

$$\beta_{TE}^2 = \frac{\mu_t}{\mu_z}\left(k_0^2\varepsilon_t\mu_z - k_t^2\right) \tag{3.23}$$

Here $k_0^2 = \omega^2\varepsilon_0\mu_0$. These two modes can be visualized from Figure 3.2, where the directions of the field vectors are shown with respect to the wave vector components.

We conclude that for any propagation direction there are two linearly polarized plane eigenwaves, one is TM and the other is TE with respect to the geometrical axis and the propagation vector, and these modes have in general different propagation constants. The only exception is the propagation along the main axis. In that case $\mathbf{k}_t = 0$, which means $\beta_{TM} = \beta_{TE}$, that is, both eigenwaves have the same propagation factors. The same is true for isotropic media, so we can say that when the wave propagates along the axis, its properties are the same as of plane waves in a simpler isotropic medium. Such directions in crystals are called *optical axes.*

The transverse fields in the eigenwaves depend on each other through wave impedances or admittances:

$$\mathbf{E}_t = \mp\overline{\overline{Z}}\cdot\mathbf{z}_0\times\mathbf{H}_t, \qquad \mathbf{z}_0\times\mathbf{H}_t = \mp\overline{\overline{Y}}\cdot\mathbf{E}_t \tag{3.24}$$

where the dyadic impedances and admittances are diagonal:

$$\overline{\overline{Z}} = Z^{TM}\frac{\mathbf{k}_t\mathbf{k}_t}{k_t^2} + Z^{TE}\frac{\mathbf{z}_0\times\mathbf{k}_t\;\mathbf{z}_0\times\mathbf{k}_t}{k_t^2}, \qquad \overline{\overline{Y}} = \overline{\overline{Z}}^{-1} \tag{3.25}$$

and the upper and lower signs correspond to waves propagating in the positive and the negative directions of the z-axis, respectively. The characteristic impedances and admittances can be found from the vector transmission-line equations (3.18) after substitution of plane linearly polarized solutions $e^{\mp j\beta_{TM}z}$ and $e^{\mp j\beta_{TE}z}$ with the eigennumbers (3.22) and (3.23). This leads to

$$Z^{TM} = \sqrt{\frac{\mu_0\mu_t}{\varepsilon_0\varepsilon_t}}\sqrt{1 - \frac{k_t^2}{k_0^2\varepsilon_z\mu_t}} \tag{3.26}$$

$$Z^{TE} = \sqrt{\frac{\mu_0\mu_t}{\varepsilon_0\varepsilon_t}}\frac{1}{\sqrt{1 - \frac{k_t^2}{k_0^2\varepsilon_t\mu_z}}} \tag{3.27}$$

As we have seen, in uniaxial media there is only one optical axis, and its direction coincides with the geometrical axis (the main axis of the permittivity tensor). In biaxial crystals, there are two optical axes, and that is the most general case. The directions of the optical axes do not in general coincide with the main axes of the permittivity tensor.

In other words, the two eigenwaves have the same propagation factors when in the plane orthogonal to the propagation direction the structure is isotropic. This can be visualized by drawing an ellipsoid *(Fresnel ellipsoid)*

$$\varepsilon_1 x^2 + \varepsilon_2 y^2 + \varepsilon_3 z^2 = \varepsilon_0 \tag{3.28}$$

If a certain plane cuts this ellipsoid so that the cross section is circular, the direction orthogonal to this plane is an optical axis. Except the sphere, every ellipsoid has at most two such different directions.

3.3 WAVEGUIDES WITH ANISOTROPIC FILLINGS

To provide a simple example of a waveguide with an anisotropic filling, let us consider again the dielectric plate waveguide, as shown in Figure 3.3. This will also demonstrate a different approach to the derivation of the dispersion relation for open waveguides. In contrast to Section 1.4.2 (Figure 1.4), the slab material is a biaxial dielectric modeled by the permittivity tensor $\overline{\overline{\varepsilon}}$ of form (3.5) and an isotropic permeability μ. The eigenvectors of the permittivity tensor are assumed to be orthogonal, and the crystal is oriented so that these axes coincide with the Cartesian axes in Figure 3.3. Thus, we can write the permittivity also as

$$\overline{\overline{\varepsilon}} = \overline{\overline{\varepsilon}}_t + \varepsilon_3 \mathbf{z}_0 \mathbf{z}_0 \tag{3.29}$$

where the "transverse permittivity" two-dimensional dyadic is

$$\overline{\overline{\varepsilon}}_t = \varepsilon_1 \mathbf{x}_0 \mathbf{x}_0 + \varepsilon_2 \mathbf{y}_0 \mathbf{y}_0 \tag{3.30}$$

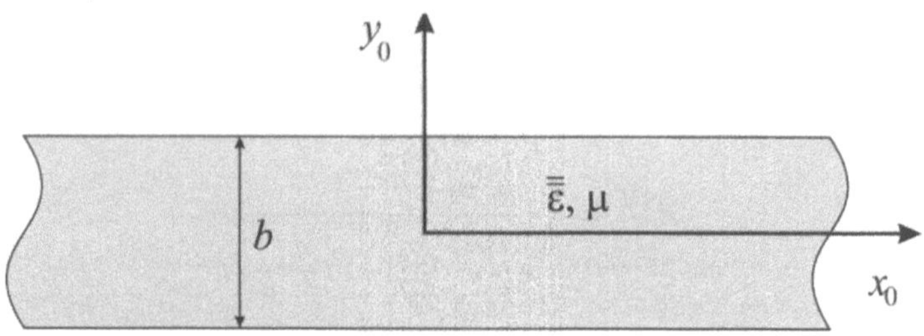

Figure 3.3. An anisotropic dielectric slab.

Let us assume that an electromagnetic wave is propagating along the $\mathbf{z}_0$ axis, therefore $\frac{\partial}{\partial z} = -jk_z$, where k_z is the longitudinal propagation constant. Also, due to the infinite dimension along the x-axis we can assume that there is no dependence on x: $\frac{\partial}{\partial x} = 0$, meaning that $\nabla = \mathbf{y}_0 \frac{\partial}{\partial y} + \mathbf{z}_0 \frac{\partial}{\partial z}$.

Starting from the Maxwell equations (1.3) we, similarly to (3.15), split the fields into longitudinal and transverse components with respect to the direction of wave propagation (axis z). The resulting equations read:

$$\frac{\partial}{\partial y} \mathbf{y}_0 \times \mathbf{E_t} + \frac{\partial}{\partial y} \mathbf{y}_0 \times \mathbf{z}_0 E_z + \frac{\partial}{\partial z} \mathbf{z_0} \times \mathbf{E}_t = -j\omega\mu \left(\mathbf{H}_t + \mathbf{z}_0 H_z \right) \tag{3.31}$$

$$\frac{\partial}{\partial y} \mathbf{y}_0 \times \mathbf{H_t} + \frac{\partial}{\partial y} \mathbf{y}_0 \times \mathbf{z}_0 H_z + \frac{\partial}{\partial z} \mathbf{z_0} \times \mathbf{H}_t = j\omega\overline{\overline{\varepsilon}} \left(\mathbf{E}_t + \mathbf{z}_0 E_z \right) \tag{3.32}$$

Writing the transversal components of these equations separately, we obtain:

$$\frac{\partial}{\partial y} \mathbf{y}_0 \times \mathbf{z}_0 E_z + \frac{\partial}{\partial z} \mathbf{z}_0 \times \mathbf{E}_t = -j\omega\mu \mathbf{H}_t \tag{3.33}$$

and similarly

$$\frac{\partial}{\partial y} \mathbf{y}_0 \times \mathbf{z}_0 H_z + \frac{\partial}{\partial z} \mathbf{z}_0 \times \mathbf{H}_t = j\omega\overline{\overline{\varepsilon}}_t \cdot \mathbf{E}_t = j\omega \left(\varepsilon_1 E_x \mathbf{x}_0 + \varepsilon_2 E_y \mathbf{y}_0 \right) \tag{3.34}$$

Let us next simplify equations (3.33) and (3.34), cross-multiplying their left- and right-hand sides by $\mathbf{z}_0$ from the left and recalling that $\frac{\partial}{\partial z} = -jk_z$:

$$\frac{\partial}{\partial y}\mathbf{y}_0 E_z + jk_z\mathbf{E}_t = -j\omega\mu\mathbf{z}_0 \times \mathbf{H}_t \tag{3.35}$$

$$\frac{\partial}{\partial y}\mathbf{y}_0 H_z + jk_z\mathbf{H}_t = j\omega\mathbf{z}_0 \times \overline{\overline{\varepsilon}}_t \cdot \mathbf{E}_t \tag{3.36}$$

Using these equations we express the x-components through the longitudinal ones:

$$H_x = \frac{j\omega\varepsilon_2}{\omega^2\varepsilon_2\mu - k_z^2}\frac{\partial}{\partial y}E_z, \qquad E_x = \frac{j\omega\mu}{-\omega^2\varepsilon_1\mu + k_z^2}\frac{\partial}{\partial y}H_z \tag{3.37}$$

The y-components are not essential when calculating the propagation constants, but we can write them, too:

$$H_y = \frac{jk_z}{\omega^2\varepsilon_1\mu - k_z^2}\frac{\partial}{\partial y}H_z, \qquad E_y = \frac{-jk_z}{\omega^2\varepsilon_2\mu - k_z^2}\frac{\partial}{\partial y}E_z \tag{3.38}$$

One can notice that the component H_x does not depend on H_z, while E_x does not depend on E_z, that is, TE and TM modes are possible.

For the TM modes let us assume that the field in the dielectric plate is distributed as a cosine function inside and exponentially decays outside:

$$E_z = \begin{cases} M_1 \cos\left(k_y y + \alpha\right) & , \quad 0 < y < \frac{b}{2} \\ M_2 \exp\left(-k_{y2} y\right) & , \quad y > \frac{b}{2} \end{cases} \tag{3.39}$$

where M_1 and M_2 are amplitude constants, k_y is the transverse propagation constant inside the anisotropic dielectric, $k_{y2} > 0$ is the decay factor, and α is a constant. Due to the symmetry, it is enough to consider only $y \geq 0$. In this case the values of α can be 0 or $\frac{\pi}{2}$, which correspond to symmetric or antisymmetric field distributions, respectively.

Applying the boundary condition for E_z at $y = \frac{b}{2}$ we obtain:

$$M_2 = M_1 \cos\left(k_y\frac{b}{2} + \alpha\right)\exp\left(k_{y2}\frac{b}{2}\right) \tag{3.40}$$

The component H_x is determined by (3.37):

$$H_x = \begin{cases} -M_1 k_y \frac{j\omega\varepsilon_2}{\omega^2\mu\varepsilon_2 - k_z^2}\sin\left(k_y\frac{b}{2} + \alpha\right) & , \quad 0 < y < \frac{b}{2} \\ -M_2 k_{y2}\frac{j\omega\varepsilon_0}{\omega^2\mu_0\varepsilon_0 - k_z^2}\exp\left(-k_{y2} y\right) & , \quad y > \frac{b}{2} \end{cases} \tag{3.41}$$

Applying the boundary condition for H_x, we make use of (3.22) (with $\varepsilon_t = \varepsilon_2$, $\varepsilon_z = \varepsilon_3$, $\beta_{TM} = k_z$, $k_t = k_y$), take into account (3.40) and substitute $\frac{\partial^2}{\partial y^2} = -k_y^2$. This results in the following equation:

$$\tan\left(k_y \frac{b}{2} + \alpha\right) = -\frac{\varepsilon_0}{\varepsilon_3}\frac{k_y}{k_{y2}} \tag{3.42}$$

which can be transformed to

$$k_y b = \pi m - 2\arctan\left(\frac{\varepsilon_0}{\varepsilon_3}\frac{k_y}{k_{y2}}\right) \tag{3.43}$$

where k_{y2} can be calculated from $k_{y2}^2 = k_z^2 - k_0^2$, and $m = 1, 2, 3\ldots$

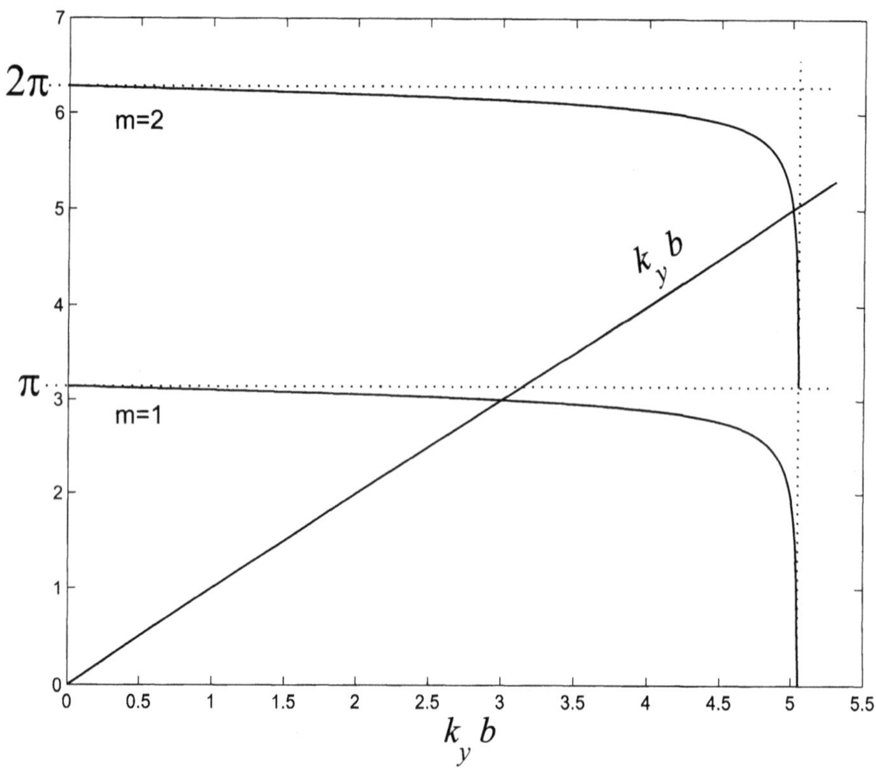

Figure 3.4. The graphical solution of equation (3.43).

The solution of (3.43) (TM modes) can be presented graphically as in Figure 3.4. Along the horizontal axis there are the values of $k_y b$, and along the vertical axis we plot the values of the expressions in the left- and right-hand sides of the dispersion equation for different values of the parameter m. As a practical example, sapphire is considered as the filling material ($\frac{\varepsilon_3}{\varepsilon_0} = 11.56$, $\frac{\varepsilon_2}{\varepsilon_0} = 9.39$), the thickness b is taken to be 1 mm, the value of $k_0 \approx 1.57$ mm^{-1} (75 GHz). As one can see, the values of $k_y b$ are limited by the value of $k_0 b\sqrt{\frac{\varepsilon_3}{\varepsilon_0} - \frac{\varepsilon_3}{\varepsilon_2}}$, as $k_{y2}^2 = k_0^2(\frac{\varepsilon_2}{\varepsilon_0} - 1) - \frac{\varepsilon_2}{\varepsilon_3} k_y^2$ cannot be smaller than zero, and the curves lay within vertical limits between 0 and π, π and 2π, etc. We can observe that two TM modes propagate in this case ($m = 1, 2$), which can be named as E_m^y, which means that the main electric field component (along the y-axis) has m extrema of the field distribution inside the slab.

For the TE modes the dispersion equation can be obtained similarly:

$$k_y b = \pi n - 2 \arctan\left(\frac{\mu_0}{\mu} \frac{k_y}{k_{y2}}\right) \tag{3.44}$$

where $n = 1, 2, 3 \ldots$, and ε_1 will play the role of ε_2.

In the case of an arbitrary orientation of the optical axes with respect to the waveguide axes TM and TE modes might not exist [4], and the calculations become more complicated. In Chapter 5 another approach is described, that is based on the exact averaging method. Furthermore, dielectric rod waveguides are considered in that chapter.

References

[1] L. Onsager, Reciprocal relations in irreversible processes, *Phys. Rev.*, vol. 37, pp. 405-426, 1931.

[2] H.B.G. Casimir, On Onsager's principle of microscopic reversibility, *Rev. Mod. Phys.*, vol. 17, pp. 343-350, 1945.

[3] L.D. Landau and E.M. Lifshitz, *Statistical Physics*, part 1, vol. 5 of *Course of Theoretical Physics*, p. 366.

[4] A.M. Goncharenko, Electromagnetic properties of a plane anisotropic waveguide, *J. of Technical Physics*, vol. 37, no. 5, 1967, pp. 822-826.

Chapter 4

Nonreciprocal media, waves in ferrite waveguides

Here we consider nonreciprocal materials (especially magnetized ferrites), which are used in microwave and millimeter-wave isolators, circulators, and polarization transformers.

4.1 PROPERTIES OF MAGNETIZED FERRITES

Electromagnetic parameters of nonreciprocal media depend on a time-odd external field. Most common practical examples are magnetized ferrites and magnetized plasmas. Ferrite inclusions are used in many microwave and millimeter-wave devices, such as isolators, circulators, tunable filters. Understanding of magnetized plasmas is important for problems of wave propagation in ionosphere. Material tensors (permittivity and permeability) of nonreciprocal materials contain antisymmetric parts. In general, the following symmetry relation is valid for linear media (the Onsager-Casimir relation [1]):

$$\overline{\overline{\varepsilon}}(\mathbf{H}_0) = \overline{\overline{\varepsilon}}^T(-\mathbf{H}_0), \qquad \overline{\overline{\mu}}(\mathbf{H}_0) = \overline{\overline{\mu}}^T(-\mathbf{H}_0) \tag{4.1}$$

where $\mathbf{H}_0$ is the external (or *bias*) magnetic field. Here we will consider microwave ferrites, which have an antisymmetric part only in the permeability tensor.

Ferrimagnetic materials are crystals with spontaneous magnetization, caused by exchange interaction between electrons in their atoms. Magnetic properties of ferrites are mainly determined by noncompensated spins of electrons. To explain these phenomena in a simple way, we will use a quasi-classical analogy. Imagine that an atom with magnetic moment $\mathbf{m}$ is positioned in an external constant magnetic field $\mathbf{H}_0$ which makes a certain angle with the direction of $\mathbf{m}$, see Figure 4.1.

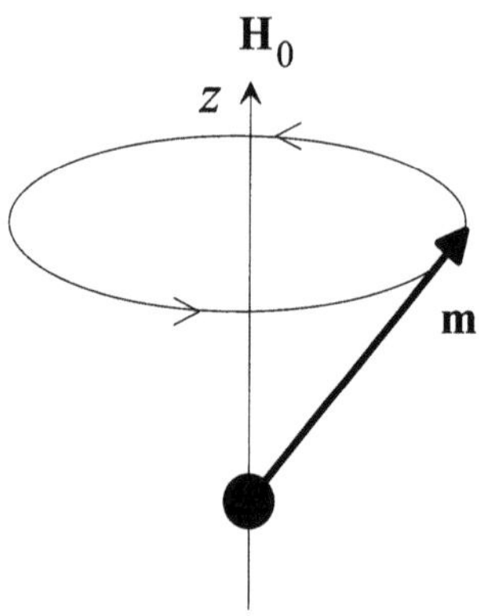

Figure 4.1. Magnetic moment ($\mathbf{m}$) precession in a constant magnetic bias field $\mathbf{H}_0$.

The momentum of force $\mathbf{T}$ acting on the atom is

$$\mathbf{T} = \mathbf{m} \times \mathbf{H}_0 \tag{4.2}$$

Assume that the movements of the atom are governed by the classical mechanical equation

$$\frac{d\mathbf{L}}{dt} = \mathbf{T} \tag{4.3}$$

where $\mathbf{L}$ is the mechanical momentum of the atom. From the quantum mechanics we know that the magnetic moment of an atom is connected with its mechanical momentum $\mathbf{L}$ as $\mathbf{m} = -\gamma \mathbf{L}$, where γ is the *gyromagnetic ratio.* For most microwave ferrites $\gamma \approx \mu_0 |e| / m_e$ (e is the electron charge and m_e is the electron mass). Thus, the equation for the magnetic moment $\mathbf{m}$ is

$$\frac{d\mathbf{m}}{dt} = -\gamma\, \mathbf{m} \times \mathbf{H}_0 \tag{4.4}$$

Multiplying by the number of magnetic atoms in the unit volume, we finally have for the magnetization vector $\mathbf{M}$ (the magnetic moment of a unit volume) the same equation:

$$\frac{d\mathbf{M}}{dt} = -\gamma\, \mathbf{M} \times \mathbf{H}_0 \tag{4.5}$$

Solutions of this equation are rotations of vector $\mathbf{M}$ around the direction of $\mathbf{H}_0$. Indeed, let us consider the time-harmonic regime and write equation (4.5) in components:

$$j\omega(M_x \mathbf{x}_0 + M_y \mathbf{y}_0) = -\gamma(-M_x H_0 \mathbf{y}_0 + M_y H_0 \mathbf{x}_0) \tag{4.6}$$

or

$$j\omega M_x = -\gamma H_0 M_y, \qquad j\omega M_y = \gamma H_0 M_x \tag{4.7}$$

Equating the determinant to zero we get

$$\omega = \gamma H_0 = \omega_H \tag{4.8}$$

Thus, the eigensolutions are oscillations at the frequency $\omega = \omega_H$ which is proportional to the bias field H_0. Substitution of this frequency value in (4.7) determines the polarization of the magnetization vector:

$$M_x = jM_y \tag{4.9}$$

That is, the vector is circularly polarized in the plane orthogonal to $\mathbf{z}_0$, as shown in Figure 4.1. Applying a circularly polarized external high-frequency magnetic field at frequencies close to ω_H, we should expect a resonant response of the material.

Let us consider the excitation of magnetization oscillations by microwave fields in detail, in order to find the permeability tensor. In our situation, the applied magnetic field is the sum of the constant bias field $\mathbf{H}_0$ and the high-frequency magnetic field $\mathbf{H}_\sim$. The amplitude of the microwave field is usually much smaller than the bias field: $H_\sim \ll H_0$. Also, the magnetization vector $\mathbf{M} = \mathbf{M}_0 + \mathbf{M}_\sim$, and $M_\sim \ll M_0$. Our goal is to solve for $\mathbf{M}_\sim$. We substitute $\mathbf{M}$ and $\mathbf{H}$ in (4.5) and *linearize* the equation, using the fact that the time-dependent components are small:

$$\frac{d\mathbf{M}_\sim}{dt} = -\gamma\left(\mathbf{M}_\sim \times \mathbf{H}_0 + \mathbf{M}_0 \times \mathbf{H}_\sim\right) \tag{4.10}$$

The second-order term $\mathbf{M}_\sim \times \mathbf{H}_\sim$ has been neglected. In components, for time-harmonic fields

$$j\omega M_x = -\gamma\left(M_y H_0 - H_y M_0\right) \tag{4.11}$$

$$j\omega M_y = -\gamma\left(H_x M_0 - M_x H_0\right) \tag{4.12}$$

$$j\omega M_z = 0 \tag{4.13}$$

(we drop the index $\sim$ for simplicity of notations). The solution for the time-harmonic component of the magnetization vector reads

$$M_x = \mu_0 \frac{\omega_M \omega_H}{\omega_H^2 - \omega^2} H_x + j\mu_0 \frac{\omega \omega_H}{\omega_H^2 - \omega^2} H_y \tag{4.14}$$

$$M_y = -j\mu_0 \frac{\omega \omega_H}{\omega_H^2 - \omega^2} H_x + \mu_0 \frac{\omega_M \omega_H}{\omega_H^2 - \omega^2} H_y \tag{4.15}$$

and, of course, $M_z = 0$. Here we have denoted $\omega_M = \gamma M_0/\mu_0$. As expected, we see that there is a resonance at $\omega = \omega_H$. The same result can be written in matrix form:

$$\mathbf{M} = \mu_0 \overline{\overline{\chi}} \cdot \mathbf{H}, \qquad \text{where} \quad \overline{\overline{\chi}} = \begin{pmatrix} \chi & j\chi_a & 0 \\ -j\chi_a & \chi & 0 \\ 0 & 0 & 0 \end{pmatrix} \tag{4.16}$$

with

$$\chi = \frac{\omega_M \omega_H}{\omega_H^2 - \omega^2}, \qquad \chi_a = \frac{\omega \omega_M}{\omega_H^2 - \omega^2}$$

Finally,

$$\mathbf{B} = \mu_0 \mathbf{H} + \mathbf{M} = \mu_0 (\overline{\overline{I}} + \overline{\overline{\chi}}) \cdot \mathbf{H} = \overline{\overline{\mu}} \cdot \mathbf{H} \tag{4.17}$$

or

$$\overline{\overline{\mu}} = \mu_0 \begin{pmatrix} \mu & j\mu_a & 0 \\ -j\mu_a & \mu & 0 \\ 0 & 0 & 1 \end{pmatrix} \tag{4.18}$$

with

$$\mu = 1 + \frac{\omega_M \omega_H}{\omega_H^2 - \omega^2}, \qquad \mu_a = \frac{\omega \omega_M}{\omega_H^2 - \omega^2} \tag{4.19}$$

In dyadic form this can be expressed as

$$\overline{\overline{\mu}} = \mu_0 (\mu \overline{\overline{I}}_t + \mathbf{z}_0 \mathbf{z}_0 - j\mu_a \mathbf{z}_0 \times \overline{\overline{I}}) \tag{4.20}$$

The unit vector $\mathbf{z}_0$ is directed along the bias field.

To account for dissipation losses, the resonant frequency can be replaced by a complex number: $\omega_H = \omega'_H - j\Delta\omega_H$. The components of the permeability tensor have the resonant frequency behavior, see Figure 4.2. Obvioulsy, the resonant frequency is proportional to the applied bias field, thus, properties of ferrites are electrically tunable.

In most millimeter-wave applications, and especially in phase shifters, ferrite samples are not magnetized to saturation, that is, the averaged static magnetization M is smaller than M_0. This is because very high

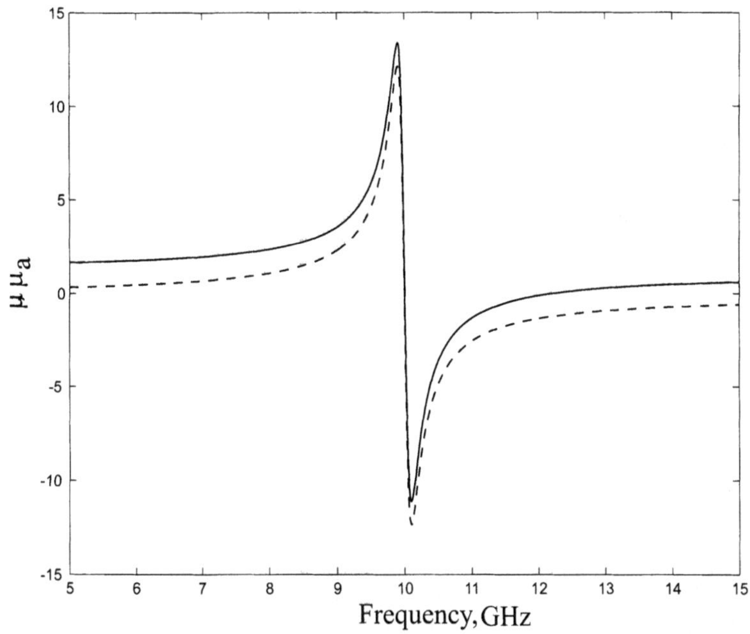

Figure 4.2. Typical frequency behavior of the two components of the permeability tensor. In this example $\omega_M/(2\pi) = 4.9$ GHz (yttrium iron garnett $Y_3Fe_5O_{12}$), and the resonant frequency $\omega_H/(2\pi) = 10$ GHz. A small loss factor was introduced. Note the resonant behavior: far from the resonance $\mu \to 1$ and $\mu_a \to 0$. Note also that near the resonance μ and μ_a can take negative values.

bias fields would be necessary to reach the ferromagnetic resonance at millimiter waves. The properties of saturated ferrites far from the resonance weakly depend on the bias field. As a result, it is only possible to electrically control the medium parameters if the bias field is below the saturation level. In this case the effective permeabity matrix retains its stucture (4.18), if the coordinate axis z is directed along the averaged magnetization vector $\mathbf{M}$, but the longitudinal component is not equal to unity:

$$\overline{\overline{\mu}} = \mu_0 \begin{pmatrix} \mu & j\mu_a & 0 \\ -j\mu_a & \mu & 0 \\ 0 & 0 & \mu_z \end{pmatrix} \tag{4.21}$$

All of these parameters are complex numbers to account for losses in the material: $\mu = \mu' - j\mu''$, $\mu_a = \mu_a' - j\mu_a''$, $\mu_z = \mu_z' - j\mu_z''$.

For weakly magnetized samples the off-diagonal term can be approximated as [2]

$$\mu'_a \approx -\frac{\gamma M}{\omega \mu_0} \tag{4.22}$$

Note that this is actually the same as the formula from (4.19) for $\omega_H \ll \omega$ with the saturation magnetization M_0 replaced by the averaged magnetization M. The transverse diagonal component can be estimated using an empirical formula [3]

$$\mu' \approx \mu'_0 + (1 - \mu'_0)\left(\frac{M}{M_0}\right)^{3/2} \tag{4.23}$$

where

$$\mu'_0 = \frac{2}{3}\left[1 - \left(\frac{\gamma M_0}{\omega \mu_0}\right)^2\right]^{1/2} + \frac{1}{3} \tag{4.24}$$

is the permeability in the completely demagnetized state [4]. At millimeter-wave frequencies we usually have $\omega \gg \omega_M$, and, as is seen from the above formula, μ'_0 is slightly smaller than unity.

For the longitudinal component μ_z there is an empirical formula [3]

$$\mu'_z \approx (\mu'_0)^\delta \tag{4.25}$$

where

$$\delta = 1 - \left(\frac{M}{M_0}\right)^{5/2} \tag{4.26}$$

The imaginary parts can be neglected if $\gamma M_0/(\omega\mu_0) \ll 1$. The loss factor greatly increases if this ratio is close to unity. More detailed theoretical analysis of partially magnetized ferrites and comparisons with experimental data can be found in [5].

4.2 LONGITUDINAL PROPAGATION. FARADAY EFFECT

Let us next consider plane waves traveling along the magnetization axis z. Since there is no dependence on x and y, the curls of $\mathbf{E}$ and $\mathbf{H}$ have no z-component:

$$\nabla \times \mathbf{E} = -\mathbf{x}_0 \frac{\partial E_y}{\partial z} + \mathbf{y}_0 \frac{\partial E_x}{\partial z} \tag{4.27}$$

Writing the Maxwell equations in components we have

$$-\frac{\partial E_y}{\partial z} = -j\omega\mu_0(\mu H_x + j\mu_a H_y) \tag{4.28}$$

$$\frac{\partial E_x}{\partial z} = -j\omega\mu_0(-j\mu_a H_x + \mu H_y) \tag{4.29}$$

$$0 = -j\omega\mu_0 H_z \tag{4.30}$$

$$-\frac{\partial H_y}{\partial z} = j\omega\varepsilon E_x \tag{4.31}$$

$$\frac{\partial H_x}{\partial z} = j\omega\varepsilon E_y \tag{4.32}$$

$$0 = j\omega\varepsilon E_z \tag{4.33}$$

First, we see that the wave is transverse: $E_z = 0$ and $H_z = 0$. Let us look for wave solutions in form $e^{-jk_z z}$. Using (4.31) and (4.32) we can eliminate the electric field components:

$$E_x = \frac{k_z}{\omega\varepsilon} H_y, \qquad E_y = -\frac{k_z}{\omega\varepsilon} H_x \tag{4.34}$$

Substituting in (4.28) and (4.29) we get

$$\left(\mu - \frac{k_z^2}{k^2}\right) H_x + j\mu_a H_y = 0 \tag{4.35}$$

$$-j\mu_a H_x + \left(\mu - \frac{k_z^2}{k^2}\right) H_y = 0 \tag{4.36}$$

where $k^2 = \omega^2\varepsilon\mu_0$. For nontrivial solutions, the determinant of this system should be zero:

$$\left(\mu - \frac{k_z^2}{k^2}\right)^2 - \mu_a^2 = 0 \tag{4.37}$$

which determines the propagation factors:

$$k_z^2 = k^2(\mu \pm \mu_a) \tag{4.38}$$

We conclude that there are two eigenwaves with different propagation factors $k_{z+} = k\sqrt{\mu + \mu_a}$ and $k_{z-} = k\sqrt{\mu - \mu_a}$. Of course, also $k_{z+} = -k\sqrt{\mu + \mu_a}$ and $k_{z-} = -k\sqrt{\mu - \mu_a}$ are solutions: They correspond to waves traveling in the opposite direction of the z-axis. Let us see what are the polarizations of the eigenwaves. To do this, we substitute k_z^2 from (4.38) into (4.35) and (4.36), which leaves us with

$$H_x = \pm j H_y \tag{4.39}$$

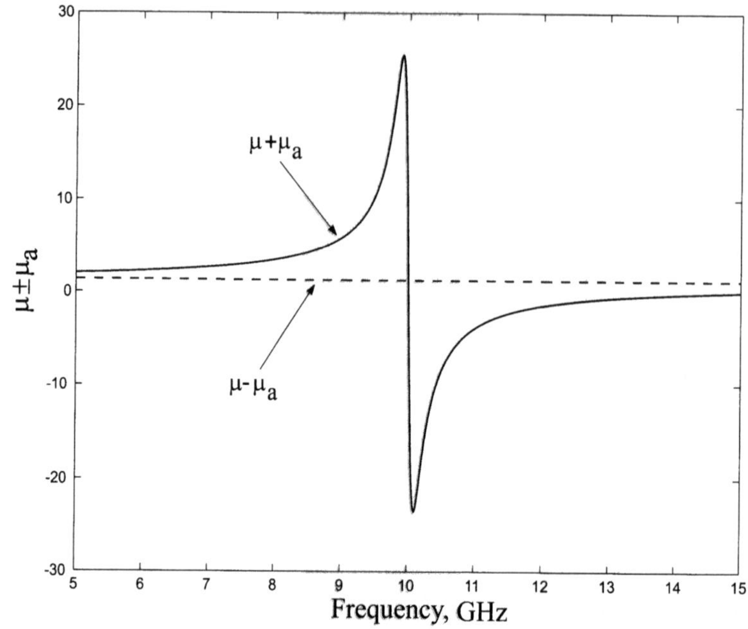

Figure 4.3. Frequency behavior of the effective permeabilities for the two eigenwaves traveling along the bias field. The same parameters as in Figure 4.2. Only one of the quantities has a resonance.

where the upper and lower signs correspond to k_{z+} and k_{z-}, respectively. Thus, the magnetic fields of the eigenwaves are circularly polarized, with the opposite directions of rotation for the two eigenwaves.

Only one of these propagation factors has a resonance, as is easy to see writing $\mu + \mu_a$ and $\mu - \mu_a$:

$$\mu + \mu_a = \frac{\omega_M + \omega_H - \omega}{\omega_H - \omega} \tag{4.40}$$

$$\mu - \mu_a = \frac{\omega_M + \omega_H + \omega}{\omega_H + \omega} \tag{4.41}$$

see also an illustration in Figure 4.3. This is because the "eigenmode" of the magnetization oscillations is circularly polarized (Section 4.1), exactly as the wave traveling with the propagation factor k_{z+}. The polarization of the other wave is orthogonal to the magnetization eigenmode, and magnetization precession is not excited by that wave. The result is that near the resonance one of the waves has a very high propagation constant and a high loss factor, but the propagation of the other wave is affected only very weakly by the ferrite filling.

Linearly polarized waves change their polarization state when they propagate in a ferrite sample along the axis, because they are not eigenwaves. A linearly polarized wave can be represented as a sum of two circularly polarized waves. Suppose that at $z = 0$ we have a wave whose magnetic field is linearly polarized and directed along x (in other words, let us direct the axis x along the magnetic field as $z = 0$). Then

$$\mathbf{H}(z) = \text{Const}\left[(\mathbf{x}_0 - j\mathbf{y}_0)e^{-jk_{z-}z} + (\mathbf{x}_0 + j\mathbf{y}_0)e^{-jk_{z+}z}\right]$$

$$= 2\text{Const}\left[\mathbf{x}_0\cos(\Delta k_z z) + \mathbf{y}_0\sin(\Delta k_z z)\right]e^{-j\overline{k_z}z} \tag{4.42}$$

where $\overline{k_z} = (k_{z+} + k_{z-})/2$ and $\Delta k_z = (k_{z+} - k_{z-})/2$. Obviously, the polarization of the magnetic field vector remains linear[1], but the polarization plane rotates with advancing z, and the rotation angle is

$$\phi(z) = \arctan(H_y/H_x) = \Delta k_z z \tag{4.43}$$

This phenomenon is called *the Faraday effect.* The rotation direction reverses if the direction of the bias field is reversed, because in that case $\mu_a \to -\mu_a$, and $k_{z\pm} \to k_{z\mp}$. At very high frequencies and for non-saturated ferrite samples, the difference between k_{z+} and k_{z-} decreases as $1/\omega$, as follows from (4.22). But the electrical thickness of the sample increases proportionally to the frequency, and the total rotation angle for a fixed length z does not vanish even at very high frequencies. With the use of the Faraday effect, it is possible to realize microwave isolators and some other devices.

In ferrite-filled waveguides magnetized along the waveguide axis, the polarization of the eigenwaves is not circular (with an exception of some modes in circular waveguides). The main effect here is the dependence of the propagation factor on the bias field, which is used in phase shifters. On microwave frequencies, metal waveguides with ferrite rods or plates are used. At millimeter-wave frequencies, open waveguides made of ferrite materials can be more practical. In the design of phase shifters, the change of μ in the process of magnetization of a ferrite element by comparatively small magnetic fields is often used. In the case of low bias fields (well below the saturation level), the antisymmetric part of the permeability tensor is small and sometimes can be neglected, but the longitudinal z-component of $\overline{\overline{\mu}}$ is not equal to μ_0 (see Section 4.1). One practical example of millimeter-wave phase shifters is shown in Figure 4.4.

[1]The electric field of this wave is elliptically polarized.

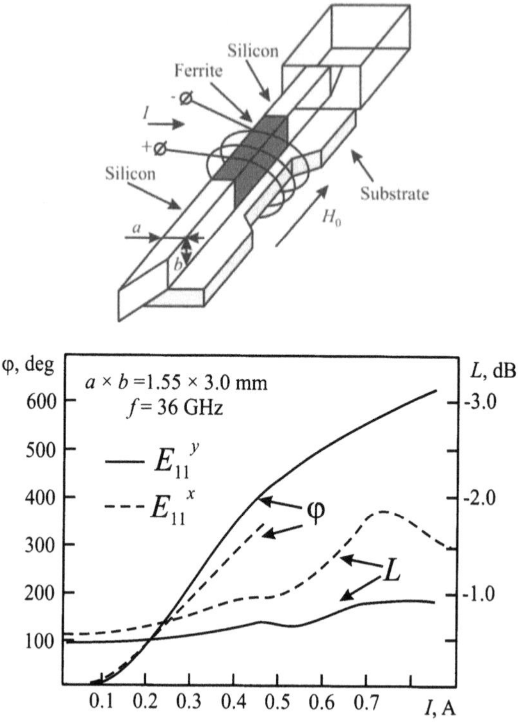

Figure 4.4. An example of a millimeter-wave ferrite phase shifter and its characteristics, from [6].

4.3 TRANSVERSE PROPAGATION

Consider next how plane waves propagate in magnetized ferrites in the direction orthogonal to the bias field. Let us assume that the permeability matrix is given by (4.18) and (4.19), that is, that the bias field is directed along z, and it is strong enough to saturate the ferrite material. Suppose that a plane wave propagates along x, and there is no field dependence along y and z. In this case the Maxwell equations read [compare with (4.28)–(4.33)]:

$$0 = -j\omega\mu_0(\mu H_x + j\mu_a H_y) \tag{4.44}$$

$$-\frac{\partial E_z}{\partial x} = -j\omega\mu_0(-j\mu_a H_x + \mu H_y) \tag{4.45}$$

$$\frac{\partial E_y}{\partial x} = -j\omega\mu_0 H_z \tag{4.46}$$

$$0 = j\omega\varepsilon E_x \tag{4.47}$$

$$-\frac{\partial H_z}{\partial x} = j\omega\varepsilon E_y \tag{4.48}$$

$$\frac{\partial H_y}{\partial x} = j\omega\varepsilon E_z \tag{4.49}$$

Obviously, the wave is TM, because $E_x = 0$. One can see that this system of equations splits into two independent sets of equations: one for E_y and H_z, and the other one for E_z, H_x, and H_y. The equations for the first set are the same as in an isotropic medium with the parameters ε and μ_0: this wave does not feel the presence of the ferrite material. This is because the high-frequency magnetic field H_z, oriented along the bias field, does not induce any precession of atomic magnetic moments. This eigenmode is called the *ordinary wave.*

Considering the other mode, let us look for a traveling wave solution in form $e^{-jk_x x}$. From (4.44) it follows that $H_x = -j\frac{\mu_a}{\mu}H_y$. Eliminating H_x, we have two equations

$$jk_x E_z + j\omega\mu_0 \frac{\mu^2 - \mu_a^2}{\mu} H_y = 0 \tag{4.50}$$

$$j\omega\varepsilon E_z + jk_x H_y = 0 \tag{4.51}$$

Equating the determinant to zero, we find the propagation constant of this wave:

$$k_x = \pm\omega\sqrt{\varepsilon\mu_0}\sqrt{\frac{\mu^2 - \mu_a^2}{\mu}} \tag{4.52}$$

This solution is called the *extraordinary wave.* The propagation constant depends on parameter

$$\mu_\perp = \frac{\mu^2 - \mu_a^2}{\mu} \tag{4.53}$$

whose typical frequency dependence is illustrated in Figure 4.5.

This mode has a cut-off at the frequency where $\mu_\perp = 0$ (see Figure 4.5), that is, at

$$\omega = \omega_\perp = \sqrt{\omega_H(\omega_H + \omega_M)} \tag{4.54}$$

Near this frequency, the effective permeability for the extraordinary wave becomes large, and the losses increase, which is typical for any resonance.

For waveguiding structures with ferrite slabs or rods magnetized in the transverse direction, the main useful effect is the nonreciprocal shift of the field distribution in the cross section of the waveguide. This we will consider using a simple example of a microstrip line on a ferrite substrate.

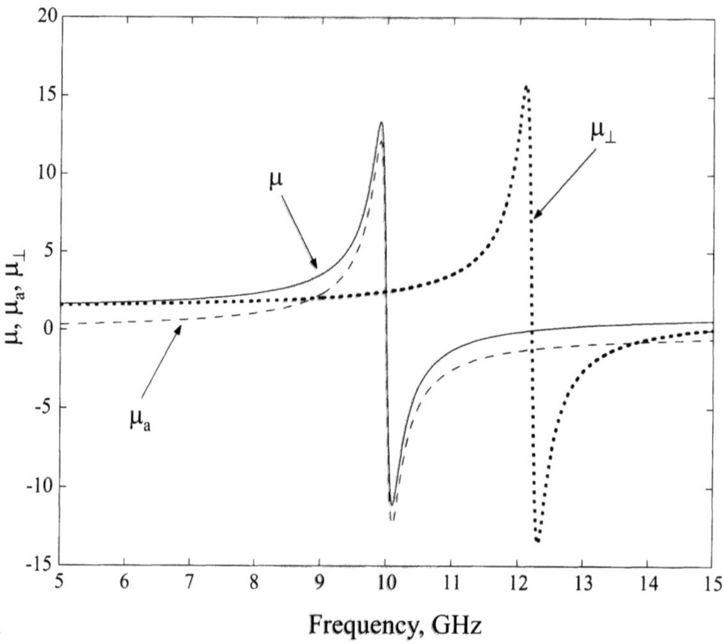

Figure 4.5. Frequency behavior of the effective permeability $\mu_\perp$ for the extraordinary wave propagating in a direction orthogonal to the bias field. Also the components of the permeability matrix μ and μ_a are shown. The last two curves are the same as in Figure 4.2.

4.3.1 Microsrip line on ferrite substrate. Isolator

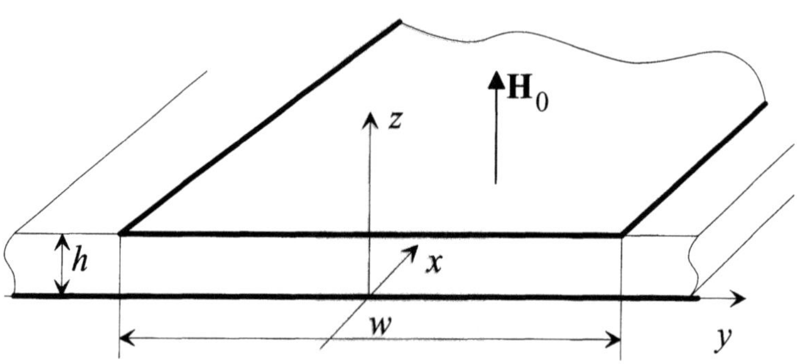

Figure 4.6. Microstrip line on a ferrite substrate. The bias magnetic field is orthogonal to the ground plane.

Microwave isolators can be realized using sections of microstrip lines on magnetized ferrite substrates. Consider a microstrip line with a wide strip, much wider than the height h ($w \gg h$), Figure 4.6. The height of the line h is much smaller than the wavelength.

The fundamental mode

In quasi-TEM waves in conventional microstrip lines, the fundamental wave is nearly transverse, that is, the field components E_z and H_y are dominant when $w \gg h$. Also in the case of ferrite substrates, it is clear that the electric field components E_x and E_y can be neglected, as compared with E_z. Indeed, these components equal zero on the ground plane and on the strip, and since the distance between the strip and the ground plane is small compared to the wavelength along the vertical direction, they are small everywhere. Let us start the analysis looking for a simple solution where only E_z and H_y are non-zero, just like in simple lines on dielectric substrates. Because the height is small, we assume that the fields are approximately uniform in the z direction.

Let us write the Maxwell equations for the fields in the waveguide. Under our assumptions

$$\nabla \times \mathbf{E} = \mathbf{x}_0 \frac{\partial E_z}{\partial y} - \mathbf{y}_0 \frac{\partial E_z}{\partial x} = -j\omega\mu_0(\mu \mathbf{y}_0 + j\mu_a \mathbf{x}_0) H_y \tag{4.55}$$

$$\nabla \times \mathbf{H} = \frac{\partial H_y}{\partial x} \mathbf{z}_0 = j\omega\varepsilon E_z \mathbf{z}_0 \tag{4.56}$$

This system splits into three scalar equations for Cartesian components. Looking for fundamental solutions in form $e^{-jk_x x - j\nu y}$, we have

$$-j\nu E_z = \omega\mu_0\mu_a H_y \tag{4.57}$$

$$jk_x E_z = -j\omega\mu_0\mu H_y \tag{4.58}$$

$$-jk_x H_y = j\omega\varepsilon E_z \tag{4.59}$$

Eliminating H_y with the use of the last equation, we get

$$k_x^2 = \omega^2 \varepsilon\mu_0\mu, \qquad \nu = -j\frac{\omega^2\varepsilon\mu_0\mu_a}{k_x} \tag{4.60}$$

This result shows that indeed such a solution with the only non-zero components E_z and H_y exists, and it has very interesting properties. The field is concentrated near one of the strip edges, since it depends on y as $e^{-j\nu y}$, and the transverse propagation factor ν is *imaginary* in the lossless case. Next, we observe that the sign of ν reverses with the sign of the propagation factor k_x. Thus, the fields of waves traveling in the opposite directions are concentrated near the opposite sides of the strip. If we introduce an absorbing layer near one of the edges of the strip, this device will work as a microwave isolator. Insertion loss will be small for waves traveling in one direction and large for reflected waves. Of course, this is a nonreciprocal effect, and indeed we see that the transverse propagation factor is proportional to the antisymmetric component μ_a of the permeability tensor. The effect disappears if $\overline{\overline{\mu}}$ becomes symmetric.

Higher-order modes

Let us now drop the assumption that there is only H_y component of the magnetic field, and write the equations assuming that also H_x can be non-zero. The result is

$$-j\nu E_z = -j\omega\mu_0(\mu H_x + j\mu_a H_y) \tag{4.61}$$

$$jk_x E_z = -j\omega\mu_0(-j\mu_a H_x + \mu H_y) \tag{4.62}$$

$$j\omega\varepsilon E_z = -jk_x H_y + j\nu H_x \tag{4.63}$$

This is a homogeneous system of linear equations for three unknowns: E_z, H_x, and H_y. Equating the determinant to zero, we have

$$k_x^2 + \nu^2 = \omega^2\varepsilon\mu_0\frac{\mu^2 - \mu_a^2}{\mu} = k^2\mu_\perp \tag{4.64}$$

where we have denoted $k^2 = \omega^2\varepsilon\mu_0$. This is very similar to the case of plane waves in isotropic dielectrics, only the permeability value is modified.

Obviously, there are two solutions of (4.64) for ν which differ by sign:

$$\nu = \pm\sqrt{k^2\mu_\perp - k_x^2} \tag{4.65}$$

Thus, the general solution for H_x reads

$$H_x = \left(Ae^{-j\nu y} + Be^{j\nu y}\right)e^{-jk_x x} \tag{4.66}$$

Assume the magnetic wall boundary conditions on the strip edges

$$H_x\big|_{y=\pm w/2} = 0 \tag{4.67}$$

which means that the magnetic field is approximately orthogonal to the side openings of the line. Imposing these conditions on (4.66), we detemine the value of the transverse propagation factor:

$$\sin(\nu w) = 0 \quad \Rightarrow \quad \nu = \pm\frac{n\pi}{w}, \quad n = 1, 2, \ldots \tag{4.68}$$

($n = 0$ gives a trivial solution). Finally, the propagation factors read, from (4.64),

$$k_{xn}^2 = k^2\mu_\perp - \left(\frac{n\pi}{w}\right)^2 \tag{4.69}$$

Of course, at low enough frequencies and not too close to the resonance (when $k\sqrt{\mu_\perp} < \pi/w$) these values are negative and the higher-order waves are evanescent.

References

[1] H.B.G. Casimir, On Onsager's principle of microscopic reversibility, *Rev. Mod. Phys.*, vol. 17, 1945, pp. 343-350.

[2] G.T. Rado, Theory of the microwave permeability tensor and Faraday effect in nonsturated ferromagnetic materials, *Phys. Rev.*, vol. 89, 1953, p. 529.

[3] J.J. Green and F. Sandy, Microwave characterization of partially magnetized ferrites, *IEEE Trans. Microwave Theory Techniques*, vol. 22, no. 6, 1974, pp. 641-645.

[4] E. Schlömann, Microwave behavior of partially magnetized ferrites, *J. Applied Physics*, vol. 41, 1970, p. 204.

[5] O. Gelin and K. Berthou-Pichavant, New consistent model for ferrite permeability tensor with arbitrary magnetization state, *IEEE Trans. Microwave Theory Techniques*, vol. 5, no. 8, 1997, pp. 1185-1192.

[6] V.V. Meriakri and M.P. Parkhomenko, Millimeter-wave dielectric strip waveguides made of ferrites and phase shifters based on these waveguides, *Electromagnetic Waves and Electronic Systems*, vol. 1, no. 1, 1996, pp. 91-96.

Chapter 5

Dielectric waveguides: classical methods for propagation constant calculations

Dielectric waveguides are more attractive at millimeter-wave frequencies, as compared with metal waveguides, because of their lower propagation loss, lower cost and easier fabrication. In case of the circular cross section the electromagnetic problems related to dielectric waveguides can be solved in closed form in terms of the Bessel functions. But in case of the rectangular cross section no closed-form solution exists. There is no strict analytical method suitable for the analysis of rectangular dielectric waveguides, moreover, numerical methods consume much memory and time and still do not always give accurate results, especially for dielectrics with high permittivities [1].

A good exposition of the numerical methods for calculating general millimeter-wave structures can be found e.g., in [2]. Several different numerical methods can be applied to the problem of dielectric waveguide structures, particularly to the open rectangular dielectric waveguide. The simplest but approximate approach is Marcatili's method.

We know that when the distribution of fields in the transverse plane depends only on one coordinate, like in a dielectric plate or a circular waveguide with axially symmetric fields, the solution can be found in terms of modes of TE and TM types (Section 1.4). However, in the case of an open rectangular dielectric waveguide, it becomes impossible, and the propagating modes are referred to as *hybrid modes*. As in [3,4], we will call the propagating mode E^y_{mn}, if its electric field is polarized mainly along the y-direction, and E^x_{mn}, if the strongest electrical field component points along the x-direction.

5.1 MARCATILI'S METHOD

One of the existing approximate methods for the rectangular dielectric waveguide is the Marcatili method [3]. It was developed for low ratios between the core permittivity and that of the cladding region (slightly more than 1), but it has been shown that it works well even for high permittivity ratios (around 10) in a rather wide frequency region.

In this method, the complete cross section area of the rectangular open dielectric waveguide (Figure 5.1) is divided into five regions with homogeneous but possibly different refractive indices n_ν in every region.

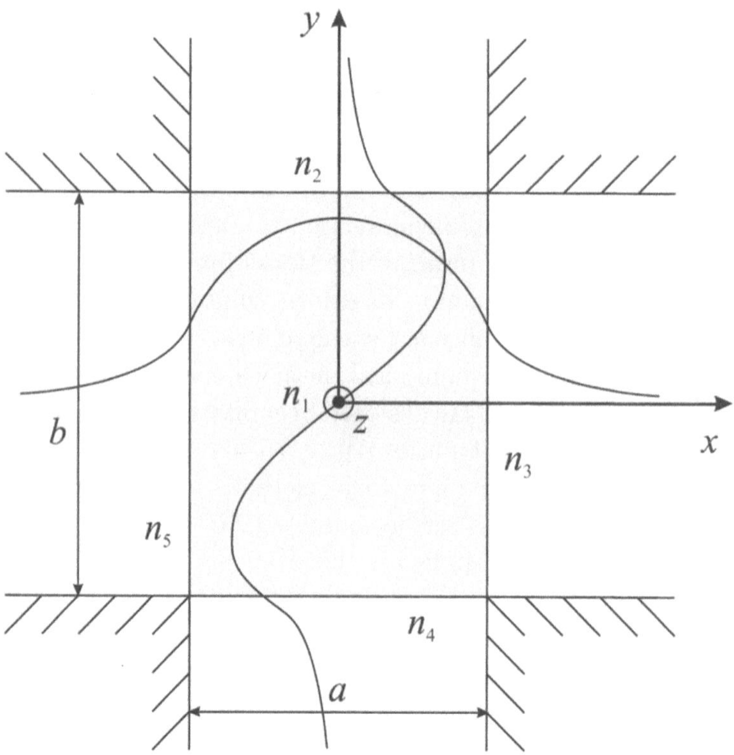

Figure 5.1. The geometry of a rectangular dielectric waveguide in Marcatili's method and the distributions of the H_x component for the E^y_{12} mode.

The fields in the shadowed regions are not considered: these regions are much less essential for the waveguide properties than the other regions. In all the other regions, the fields are assumed to be approximately (co)sinusoidally distributed inside the waveguide and decaying exponentially outside. That is, we express the field components, say, H_x, as follows:

$$H_{x\nu} = \exp(-jk_z z + j\omega t) \begin{cases} M_1 \cos(k_x x + \alpha)\cos(k_y y + \beta) & \text{for} \quad \nu = 1 \\ M_2 \cos(k_x x + \alpha)\exp(-k_{y2} y) & \text{for} \quad \nu = 2 \\ M_3 \cos(k_y y + \beta)\exp(-k_{x3} x) & \text{for} \quad \nu = 3 \\ M_4 \cos(k_x x + \alpha)\exp(k_{y4} y) & \text{for} \quad \nu = 4 \\ M_5 \cos(k_y y + \beta)\exp(-k_{x5} x) & \text{for} \quad \nu = 5 \end{cases} \tag{5.1}$$

where $M_{1...5}$ are unknown amplitude coefficients, k_x and k_y are the propagation constants in region 1 (refractive index n_1) in the horizontal and vertical directions, respectively, k_{x3} and k_{y2} are the decay factors in the outer regions, and α and β are additional phase constants. If the system is symmetric, meaning that $n_2 = n_4$ and $n_3 = n_5$, the field distributions are given by cosine or sine functions with a null or a maximum at the rod center, and in this case constants α and β equal either 0 or $\frac{\pi}{2}$. This notation allows us to write both symmetric and antisymmetric field distributions in a compact form, as in (5.1). For the orthogonal polarization we can write similar expressions for H_y, but at this stage let us assume that $H_y = 0$. The other field components can be then expressed through H_x, using (1.16) and (1.17). Finally, the field components are written as

$$H_{y\nu} = 0 \tag{5.2}$$

$$H_{z\nu} = -\frac{j}{k_z}\frac{\partial H_{x\nu}}{\partial x} \tag{5.3}$$

$$E_{x\nu} = -\frac{1}{\omega\varepsilon_0 n_\nu^2 k_z}\frac{\partial^2 H_{x\nu}}{\partial x \partial y} \tag{5.4}$$

$$E_{y\nu} = \frac{k_0^2 n_\nu^2 - k_{y\nu}^2}{\omega\varepsilon_0 n_\nu^2 k_z} H_{x\nu} \tag{5.5}$$

$$E_{z\nu} = \frac{j}{\omega\varepsilon_0 n_\nu^2}\frac{\partial H_{x\nu}}{\partial y} \tag{5.6}$$

where k_0 is the wave number in vacuum.

5.1.1 Rectangular dielectric rod waveguide in air

To analyze the open rectangular waveguide (Figure 5.1) we return to the main equations (5.1) and consider waveguides in air, with $n_{2,3,4,5} = 1$. Due to this symmetry α, $\beta = 0$ or $\pm\frac{\pi}{2}$ in (5.1). The symmetry also allows us to consider only the region $x > 0$, $y > 0$, see Figure 5.1. To find the eigensolutions we should match the components H_x, E_z, H_z, E_x at $y = \frac{b}{2}$ and E_y, E_z, H_z at $x = \frac{a}{2}$. In his original paper [3] Marcatili assumed that the refractive index of the dielectric waveguide was only slightly larger than that of air: $n_1 - 1 \ll 1$, therefore $k_{x,y} \ll k_z$, because the fields are more spread out from the rod when the dielectric material properties approach the properties of air. Thus, E_x can be neglected, as it is proportional to $k_x k_y$. Now, let us match the strongest field component, E_y, at $x = \frac{a}{2}$. Neglecting k_y^2 as compared with $k_0^2 n_1^2$, we come to

$$\frac{k_0^2}{\omega\varepsilon_0 k_z} M_1 \cos\left(k_x \frac{a}{2} + \alpha\right) = \frac{k_0^2}{\omega\varepsilon_0 k_z} M_3 \exp\left(-k_{x3}\frac{a}{2}\right) \tag{5.7}$$

From here,

$$M_3 = M_1 \cos\left(k_x \frac{a}{2} + \alpha\right) \exp\left(k_{x3}\frac{a}{2}\right) \tag{5.8}$$

When we write the boundary condition for E_z at $x = \frac{a}{2}$ and solve the resulting equation, we do not come to any useful result, because from

$$-\frac{j}{\omega\varepsilon_0 n_1^2} M_1 k_y \cos\left(k_x \frac{a}{2} + \alpha\right) = -\frac{j}{\omega\varepsilon_0} k_y M_3 \exp\left(-k_{x3}\frac{a}{2}\right) \tag{5.9}$$

we arrive at $\frac{1}{n_1^2} = 1$, which leaves us with a trivial relation. For this reason we will not take into account this boundary condition. Fortunately, the boundary condition for H_z gives

$$\frac{j}{k_z} M_1 k_x \sin\left(k_x \frac{a}{2} + \alpha\right) = \frac{j}{k_z} M_3 k_{x3} \exp\left(-k_{x3}\frac{a}{2}\right) \tag{5.10}$$

which can be rewritten as

$$k_x \sin\left(k_x \frac{a}{2} + \alpha\right) = k_{x3} \cos\left(k_z \frac{a}{2} + \alpha\right) \tag{5.11}$$

or, after some transformations,

$$k_x a = \pi n - 2 \arctan\left(\frac{k_x}{k_{x3}}\right) \tag{5.12}$$

where $k_{x3} = \sqrt{k_0^2(n_1^2 - 1) - k_x^2}$, and $n = 1, 2, 3, \ldots$ At $y = \frac{b}{2}$ we similarly match H_x and E_z and do not consider H_z. From the boundary condition for H_x we get

$$M_1 \cos\left(k_y \frac{b}{2} + \beta\right) = M_2 \exp\left(-k_{y2} \frac{b}{2}\right) \tag{5.13}$$

$$-\frac{j}{\omega \varepsilon_0 n_1^2} k_y M_1 \sin\left(k_y \frac{b}{2} + \beta\right) = -\frac{j}{\omega \varepsilon_0} k_{y2} M_2 \exp\left(-k_{y2} \frac{b}{2}\right) \tag{5.14}$$

or

$$\frac{1}{n_1^2} k_y \sin\left(k_y \frac{b}{2} + \beta\right) = k_{y2} \cos\left(k_y \frac{b}{2} + \beta\right) \tag{5.15}$$

From here we arrive at

$$k_y b = \pi m - 2 \arctan\left(\frac{1}{n_1^2} \frac{k_y}{k_{y2}}\right) \tag{5.16}$$

where $k_{y2} = \sqrt{k_0^2(n_1^2 - 1) - k_y^2}$ and $m = 1, 2, 3, \ldots$ Thus, to calculate the propagation constant of a rectangular dielectric waveguide, we have to solve equations (5.12) and (5.16) and then find

$$k_z = \sqrt{k_0^2 n_1^2 - k_x^2 - k_y^2} \tag{5.17}$$

That is, we have to solve for two dielectric slabs (as in Section 3.3): one vertical and one horizontal, with the thicknesses a and b, respectively. Another approach can be found in [4], where all the transversal field components are expressed through the longitudinal ones for the case of arbitrary values of $n_{2,3,4,5}$ (not necessarily symmetric cladding). In the case of symmetrical waveguides the resulting equations are the same as above.

5.1.2 Some properties of rectangular dielectric waveguides

Indices m and n in the equations for the eigenmodes (5.12) and (5.16) denote how many extrema the distribution of the main field component has in the horizontal and vertical directions, respectively. For example, in Figure 5.1 the field distribution for E_{12}^y mode is shown.

In Figure 5.2 typical propagation and loss characteristics are shown. One can see that the dielectric waveguide has two fundamental modes without a low-frequency cut-off, which are degenerate if the cross section

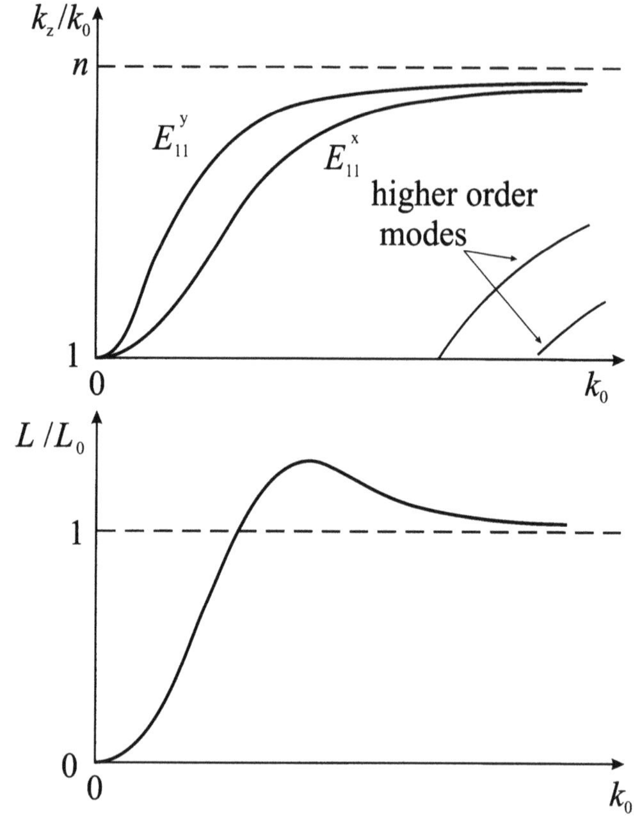

Figure 5.2. Typical dependence of the normalized propagation factor $\frac{k_z}{k_0}$ and the loss factor $\frac{L}{L_0}$ for the fundamental mode on the wave number in vacuum k_0. Here L_0 is the propagation loss of a plane wave in the infinite medium filled with the same material as the dielectric waveguide core.

is a square with $a = b$. At very high frequencies the loss factor tends to that for plane waves in the core because of the strong field concentration in the rod. At low frequencies losses are small because the fields are only weakly concentrated in the rod (for this reason, however, the usage of rod waveguides at very low frequencies is not practical).

5.1.3 How well does Marcatili's method work?

In the original Marcatili method it is assumed that the refractive index of the dielectric waveguide is only slightly larger than unity. Let us check what happens if we try to calculate the fundamental E_{11}^y mode for, say, a Silicon dielectric waveguide ($n = 3.41$) with the cross section dimensions

0.5×1.0 mm^2. Next we compare the results with the same characteristics calculated with a more accurate numerical Goell's method [5]. Later in this chapter we will explain Goell's method in more detail and extend that to the case of uniaxial anisotropic core waveguides.

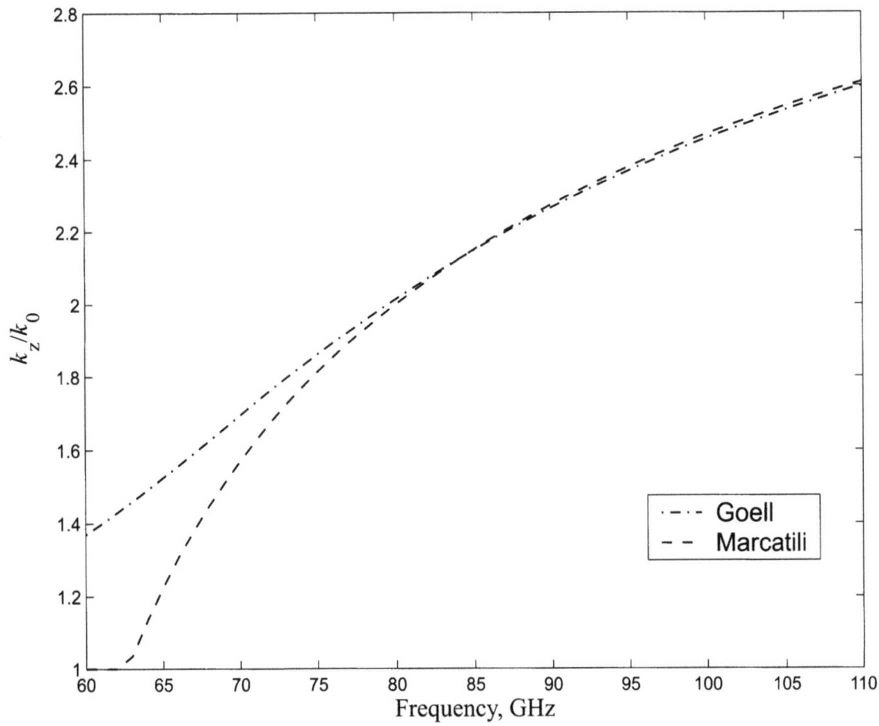

Figure 5.3. Comparison of the results of calculations of the propagation factor $\frac{k_z}{k_0}$ by Goell's and Marcatili's methods for a Silicon dielectric waveguide with 0.5×1.0 mm^2 cross section, E^y_{11} mode.

One can see from Figure 5.3 that in spite of its simplicity Marcatili's method works quite well for dielectric waveguides made of even high permittivity materials like Silicon. However, it works well only when the wave is well guided, that is, at high enough frequencies. At lower frequencies accurate calculations become more complicated [1].

5.2 GOELL'S METHOD

Instead of expanding the field distribution in sinusoidal functions, Goell [5] proposed to use the following approach: he expanded the longitudinal components of the electric and magnetic fields as

$$E_{z1} = \sum_{n=0}^{+\infty} a_n J_n(\beta_t r) \sin(n\theta + \phi_n) \exp(-jk_z z + j\omega t) \tag{5.18}$$

$$H_{z1} = \sum_{n=0}^{+\infty} b_n J_n(\beta_t r) \sin(n\theta + \psi_n) \exp(-jk_z z + j\omega t) \tag{5.19}$$

inside the core of the dielectric waveguide and

$$E_{z0} = \sum_{n=0}^{+\infty} c_n K_n(qr) \sin(n\theta + \phi_n) \exp(-jk_z z + j\omega t) \tag{5.20}$$

$$H_{z0} = \sum_{n=0}^{+\infty} d_n K_n(qr) \sin(n\theta + \psi_n) \exp(-jk_z z + j\omega t) \tag{5.21}$$

outside the core. Here $\beta_t = \sqrt{k_1^2 - k_z^2}$ and $q = \sqrt{k_z^2 - k_0^2}$ are the transverse propagation constants inside and outside of the dielectric rod, $J(\beta_t r)$ and $K(qr)$ are the Bessel functions and the modified Bessel functions of the second kind, respectively. Variables a, b, c, and d are weighting coefficients, and ϕ and ψ are phase constants. Definitions of angle θ and distance r are shown in Figure 5.4.

The sums are truncated at some finite number of terms N. Then, the tangential field components are expressed and "tailored" (requiring the boundary conditions to be satisfied) at several points of the interface. The symmetry simplifies the analysis [5], because one can consider only the first quadrant ($0 < \theta < \frac{\pi}{2}$) and only odd or even numbers of n and corresponding to them harmonics. For example, for the odd harmonic case Goell chose the matching points as:

$$\theta_m = \left(m - \frac{1}{2}\right) \frac{\pi}{2N} \tag{5.22}$$

with respect to the angle $\theta = 0$, $m = 1, 2, \ldots, N$. As a result of this "tailoring", a system of equations is obtained, for which the determinant of its matrix should be equal to zero, otherwise no nonzero solution can exist. This method is rather fast, does not require many expansion terms (we used 10-12 terms), and is considered as classical.

A more rigorous solution without the limitation of matching only at particular points of the interface is proposed in [6]. The same expansion

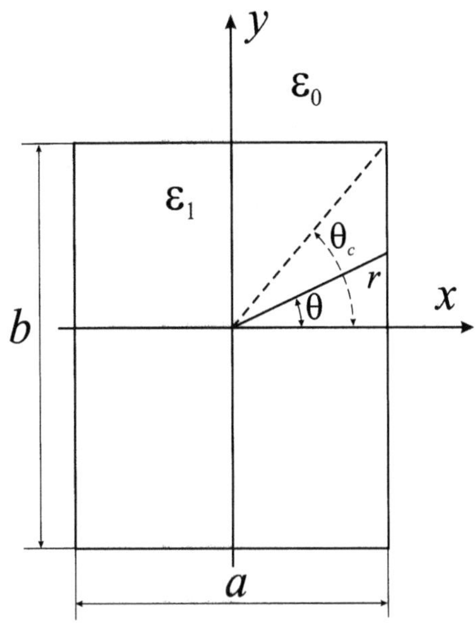

Figure 5.4. Dimensions of the dielectric waveguide cross section and the corresponding coordinate system.

of the longitudinal field components is used except that instead of sine (cosine) functions complex exponents are taken. Then, the exact boundary conditions are written in terms of the longitudinal field components and their derivatives both in the tangential and normal directions. As a result, an infinite system of equations is obtained and the determinant of the matrix is set to zero. The solution in this case is much more complicated but more rigorous. In the next section more details on Goell's method will be explained, and the method will be extended to a more general anisotropic case.

5.3 OPEN ANISOTROPIC WAVEGUIDES

5.3.1 Modification of Marcatili's method for the calculation of anisotropic rectangular dielectric waveguides

The uniaxial anisotropic dielectric rod waveguide is nowadays of increasing interest, because many high-quality dielectric materials available for fabrication of millimeter-wave waveguides are uniaxial crystals (for example, sapphire). A simple theoretical approximate model for the calculation of the propagation constant is required in addition to accurate but complicated ones (see, e.g., [7]). We have seen that Marcatili's method is quite effective for many practically important millimeter-wave rectangular dielectric waveguides, but it cannot be directly used for anisotropic cores. In this section we present an approach to the solution of the uniaxial anisotropic case, which displays good accuracy despite its simplicity [8].

The principal focus here will be the case in which the optical axis of the material coincides with the axis of the dielectric rod waveguide, in which case we can avoid excitation of the orthogonal mode and/or rotation of the polarization plane. The E_{11}^y fundamental mode of the rectangular dielectric waveguides with cross section $a \times b$ is selected, and using its symmetry we formulate the problem, as is shown in Figure 5.5.

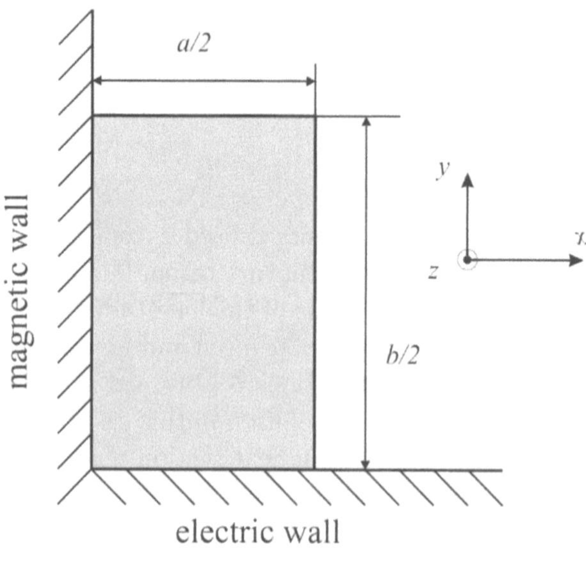

Figure 5.5. Cross section of the waveguide problem space.

To solve this problem we make use of the fact [9], that Marcatili's method for a rectangular dielectric rod waveguide with the cross section $a \times b$ actually reduces to solving two problems for two dielectric slabs: a horizontal slab of thickness b and a vertical slab of thickness a for the same polarization. Analyzing the geometry shown in Figure 5.6, we complement the classical Marcatili approach with the exact averaging method [10]. Probably the main advantage of this approach (in addition to its compactness and simplicity) is the fact that it allows one to derive the dispersion equations without guessing the field distribution.

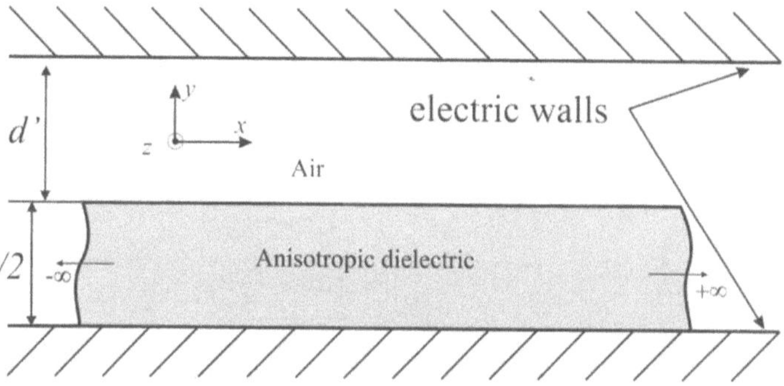

Figure 5.6. Geometry of an anisotropic dielectric slab waveguide problem.

The analysis can be simplified noticing that in the case of the vertical dielectric slab for the mode of interest there is no longitudinal component E_z, only H_z, because $\frac{\partial}{\partial y} = 0$ and $H_y = 0$ [11]. Thus, only the E_y component is present, and the corresponding equation in Marcatili's method does not need to be changed. Next, let us solve the dielectric slab problem shown in Figure 5.6, when the dielectric permittivity is determined by the matrix

$$\overline{\overline{\varepsilon}} = \begin{pmatrix} \varepsilon_\perp & 0 & 0 \\ 0 & \varepsilon_\perp & 0 \\ 0 & 0 & \varepsilon_{||} \end{pmatrix} \tag{5.23}$$

where $\varepsilon_\perp = \varepsilon_{r\perp}\varepsilon_0$ denotes the permittivity in the directions normal to the optical axis and $\varepsilon_{||} = \varepsilon_{r||}\varepsilon_0$ denotes the permittivity for the fields parallel to the optical axis. In the following, permeability μ is assumed to be a scalar. In the exact averaging method the vector transmission-line equations for the fields in the slab (1.11) and (1.12) are integrated over the slab thickness, and the averaged tangential fields are introduced as

$$\hat{\mathbf{E}}_t = \int_0^d \mathbf{E}_t \, dy, \qquad \hat{\mathbf{H}}_t = \int_0^d \mathbf{H}_t \, dy \tag{5.24}$$

For a uniaxial slab this leads to the following relations:

$$\frac{\mathbf{n} \times \mathbf{E}_{t+} - \mathbf{n} \times \mathbf{E}_{t-}}{d} = -j\omega\mu\hat{\mathbf{H}}_t - \frac{1}{j\omega\varepsilon_\perp}\nabla \times \nabla \times \hat{\mathbf{H}}_t \tag{5.25}$$

$$\frac{\mathbf{n}\times\mathbf{H}_{t+}-\mathbf{n}\times\mathbf{H}_{t-}}{d}=j\omega\varepsilon_{||}\hat{\mathbf{E}}_t+\frac{1}{j\omega\mu}\nabla\times\nabla\times\hat{\mathbf{E}}_t \tag{5.26}$$

Next, from the known general solution for the fields inside the slab (a combination of plane waves) we find the exact relation between the averaged fields and the field values on the two opposite sides of the slab [10]:

$$\hat{\mathbf{E}}_t=(\mathbf{E}_{t+}+\mathbf{E}_{t-})\,f(\beta_y d) \tag{5.27}$$

where

$$f(\beta_y d)=\frac{\tan\left(\frac{\beta_y d}{2}\right)}{\beta_y d}$$

It is assumed that the permeability μ of the slab material coincides with that of vacuum. Subscript indices $t+$ and $t-$ refer to the tangential components of the corresponding vector on the upper and lower sides of the slab, respectively. The transverse propagation constant $\beta_y = \sqrt{\frac{\varepsilon_{r||}}{\varepsilon_{r\perp}}\left(k_0^2\varepsilon_{r\perp}-k_z^2\right)}$.

In our case, $\mathbf{E}_{t-}=0$ because of the presence of an electric wall. Substituting $\nabla_t=-jk_z$ into (5.25)–(5.26) and rewriting these equations for the x- and z-components separately, we find for the z-components:

$$\frac{E_{z+}}{d}=\frac{\beta_y^2}{j\omega\varepsilon_{||}}(H_{x+}+H_{x-})\,f(\beta_y d) \tag{5.28}$$

$$\frac{H_{z+}-H_{z+}}{d}=\frac{-\omega^2\varepsilon_{||}\mu+k_z^2}{j\omega\mu}E_{x+}f(\beta_y d) \tag{5.29}$$

and for the x-components:

$$\frac{E_{x+}}{d}=j\omega\mu\,(H_{z+}+H_{z-})\,f(\beta_y d) \tag{5.30}$$

$$\frac{H_{x+}-H_{x-}}{d}=-j\omega\varepsilon_{||}E_{z+}f(\beta_y d) \tag{5.31}$$

One can notice that in (5.28) and (5.31) only H_x and E_z are present, while in (5.29) and (5.30) there are only H_z and E_x components. Let us choose the equations corresponding to the E_1^y mode, i.e., E_x and H_z are equal to zero and only (5.28) and (5.31) are nontrivial. Eliminating H_{x-} we find

$$2H_{x+}=j\omega\varepsilon_{||}E_{z+}\frac{2}{\beta_y}\frac{1}{(\beta_y d)} \tag{5.32}$$

For the air layer in Figure 5.6, one can write similarly:

$$2H_{x+} = -j\omega\varepsilon_0 E_{z+} \frac{2}{\beta'_y} \frac{1}{(\beta'_y d')} \tag{5.33}$$

where $\beta'^2_y = k_0^2 - k_z^2$. The minus sign here comes from the fact that the normal vector points in the opposite direction. When d' goes to infinity, (5.33) should be replaced by [10]

$$H_{x+} = j\omega\varepsilon_0 E_{z+} \frac{j}{\beta'_y} \tag{5.34}$$

Since the propagation constant in the structure in Figure 5.6 is larger than that in air, β'_y is an imaginary number. Combining these equations, viz., (5.32) and (5.34), we obtain:

$$\frac{\varepsilon_{r||}}{\beta_y} \frac{1}{\tan(\beta_y d)} = \frac{j}{\beta'_y} \tag{5.35}$$

In the notations of the original paper by Marcatili [3], let us write $k_y = \beta_y$, $k_{y2} = \frac{\beta'_y}{j}$, and $b = 2d$. Using the relation $\frac{1}{\tan(x)} = \tan\left(\frac{\pi}{2} - x\right)$, an equation similar to the "Marcatili's equation" [3,11] can be obtained:

$$k_y b = \pi + 2\pi m - 2\arctan\left(\frac{1}{\varepsilon_{r||}} \frac{k_y}{k_{y2}}\right) \tag{5.36}$$

where $m = 0, 1, 2, \ldots$, $k_y = \sqrt{\frac{\varepsilon_{r||}}{\varepsilon_{r\perp}}\left(k_0^2 \varepsilon_{r\perp} - k_z^2\right)}$, $k_{y2} = \sqrt{k_z^2 - k_0^2}$. The case of $m = 0$ corresponds to the fundamental E_1^y mode. To obtain the remaining possible modes (E_z is even with respect to $y = 0$ plane), one can position the slab on a magnetic wall. Regarding the vertical slab, no longitudinal electrical field component is present for the E_1^y mode, therefore the "longitudinal" permittivity does not change the second "Marcatili equation" [3]:

$$k_x a = \pi n - 2\arctan\left(\frac{k_x}{k_{x3}}\right) \tag{5.37}$$

where $n = 1, 2, 3, \ldots$ Thus, the solution procedure can be organized as follows. Using (5.36) we find k_z, then solve (5.37) for k_x, and, eventually, find the correct $k_{z\ \text{final}}$:

$$k_{z\ \text{final}} = \sqrt{k_z^2 - k_x^2} \tag{5.38}$$

The above results are valid for rectangular dielectric waveguides made of a uniaxial anisotropic dielectric material with the optical axis coinciding with the axis of the dielectric rod waveguide. In more general cases (arbitrary direction of the optical axis, anisotropic permeability, etc.) equations similar to (5.25)–(5.26) can be derived and solved using the same approach. Let us stress that the main equations have been derived without guessing on the actual field distribution. This method is relatively simple and sufficiently accurate for dielectric waveguides operating far from the cut-off, which is also the case with the original Marcatili method for isotropic waveguides.

5.3.2 Application of Goell's method for the calculation of anisotropic rectangular dielectric waveguides

Similarly to the case of waveguides with isotropic cores, more accurate results can be obtained using the Goell method. Here we will show how to use that method for anisotropic waveguides [12].

The cross-section of the dielectric waveguide under study is shown in Figure 5.4. As before, let us introduce the uniaxial permittivity matrix (5.23). Similarly to [5], we assume that the longitudinal components of the electric and magnetic fields inside the core are distributed as

$$E_{z1} = \sum_{n=0}^{\infty} a_n J_n(\beta_{TM} r) \sin(n\theta + \phi_n) \exp(-jk_z z + j\omega t) \tag{5.39}$$

$$H_{z1} = \sum_{n=0}^{\infty} b_n J_n(\beta_{TE} r) \sin(n\theta + \psi_n) \exp(-jk_z z + j\omega t) \tag{5.40}$$

inside the waveguide core, and as

$$E_{z0} = \sum_{n=0}^{\infty} c_n K_n(qr) \sin(n\theta + \phi_n) \exp(-jk_z z + j\omega t) \tag{5.41}$$

$$H_{z0} = \sum_{n=0}^{\infty} d_n K_n(qr) \sin(n\theta + \psi_n) \exp(-jk_z z + j\omega t) \tag{5.42}$$

outside the core. The main difference of these equations from those in [5] is in the presence of two transversal propagation constants β_{TM} and β_{TE}, which read [13] [see also Chapter 3, equations (3.22), (3.23)]:

$$\beta_{TE}^2 = k_0^2 \varepsilon_{r\perp} - k_z^2 \tag{5.43}$$

$$\beta_{TM}^2 = k_0^2 \varepsilon_{r||} - \frac{\varepsilon_{||}}{\varepsilon_{\perp}} k_z^2 \tag{5.44}$$

The mode of our interest is E_{11}^y. Therefore, using the symmetry of the structure, one can substitute into (5.39)–(5.42) [5] the following:

$$\phi_n = 0, \quad \psi_n = \frac{\pi}{2} \tag{5.45}$$

and consider only the angles between 0 and $\frac{\pi}{2}$. It can be shown using e.g. [13] that the equations for the transverse field components do not have to be changed but remain as follows:

$$E_r = \frac{-jk_z}{k^2 - k_z^2} \left[\frac{\partial E_z}{\partial r} + \left(\frac{\mu_0 \omega}{k_z r} \right) \frac{\partial H_z}{\partial \theta} \right] \tag{5.46}$$

$$E_\theta = \frac{-jk_z}{k^2 - k_z^2} \left[\frac{1}{r} \frac{\partial E_z}{\partial \theta} + \left(\frac{\mu_0 \omega}{k_z} \right) \frac{\partial H_z}{\partial r} \right] \tag{5.47}$$

$$H_r = \frac{-jk_z}{k^2 - k_z^2} \left[- \left(\frac{k^2}{\mu_0 \omega k_z r} \right) \frac{\partial E_z}{\partial \theta} + \frac{\partial H_z}{\partial r} \right] \tag{5.48}$$

$$H_\theta = \frac{-jk_z}{k^2 - k_z^2} \left[- \left(\frac{k^2}{\mu_0 \omega k_z} \right) \frac{\partial E_z}{\partial r} + \frac{1}{r} \frac{\partial H_z}{\partial \theta} \right] \tag{5.49}$$

where $k^2 = k_0^2 \varepsilon_{r\perp} \mu$. The tangential components of the electric field are similarly given by

$$E_t = E_r \sin(\theta) + E_\theta \cos(\theta), \qquad \text{if } \theta < \theta_c$$

$$E_t = -E_r \cos(\theta) + E_\theta \sin(\theta), \qquad \text{if } \theta > \theta_c \tag{5.50}$$

where the value of θ_c is shown in Figure 5.4.

The next step is to match the tangential field components on the interfaces. The matching points on the dielectric waveguide surface can be chosen to be at the angles

$$\theta_m = \frac{\left(m - \frac{1}{2}\right) \pi}{2N} \tag{5.51}$$

where $m = 1, 2, \ldots, N$ is the number of the point for the fields to be matched at, and N is the number of harmonics taken into account. Moreover, only odd n in equations (5.39)–(5.42) are selected [5]. After "tailoring" the tangential field components inside and outside the core of the waveguide, one can obtain an equation similar to that given in [5], but the components in it are changed to be as follows:

$$\det \begin{pmatrix} E^{LA} & 0 & -E^{LC} & 0 \\ 0 & H^{LB} & 0 & -H^{LD} \\ E^{TA} & E^{TB} & -E^{TC} & -E^{TD} \\ H^{TA} & H^{TB} & -H^{TC} & -H^{TD} \end{pmatrix} = 0 \qquad (5.52)$$

where the submatrix elements are:

$$e_{mn}^{LA} = J_{TM}S$$

$$h_{mn}^{LB} = J_{TE}C$$

$$e_{mn}^{LC} = KS$$

$$h_{mn}^{LD} = KC$$

$$e_{mn}^{TA} = -k_z\left[\mathbf{J}'_{TM}SR + \mathbf{J}_{TM}CT\right]$$

$$e_{mn}^{TB} = k_0 Z_0\left[\mathbf{J}_{TE}SR + \mathbf{J}'_{TE}CT\right]$$

$$e_{mn}^{TC} = k_z\left[\mathbf{K}'SR + \mathbf{K}CT\right]$$

$$e_{mn}^{TD} = -k_0 Z_0\left[\mathbf{K}SR + \mathbf{K}'CT\right]$$

$$h_{mn}^{TA} = \varepsilon_\perp k_0[\mathbf{J}_E CR - \mathbf{J}'_E ST]/Z_0$$

$$h_{mn}^{TB} = -k_z[\mathbf{J}'_M CR - \mathbf{J}_M ST]$$

$$h_{mn}^{TC} = -k_0[\mathbf{K}CR - \mathbf{K}'ST]/Z_0$$

$$h_{mn}^{TD} = k_z[\mathbf{K}'CR - \mathbf{K}ST]$$

One can say that the Bessel functions used here are separated into "electric" (subscript TM) and "magnetic" (subscript TE) kinds. Furthermore, we have used the following notations:

$$Z_0 = \sqrt{\frac{\mu_0}{\varepsilon_0}}$$

$$S = \sin[(2n-1)\theta_m + \phi], \quad C = \cos[(2n-1)\theta_m + \phi], \quad \phi = 0$$

$$J_{TM} = J_{2n-1}(\beta_{TM} r_m), \quad J_{TE} = J_{2n-1}(\beta_{TE} r_m), \quad K = K_{2n-1}(q r_m)$$

$$\mathbf{J}_{TM} = \frac{(2n-1)J_{2n-1}(\beta_{TM} r_m)}{\beta_{TE}^2 r_m}, \quad \mathbf{J}_{TE} = \frac{(2n-1)J_{2n-1}(\beta_{TE} r_m)}{\beta_{TE}^2 r_m}$$

$$\mathbf{J}'_{TM} = \frac{J'_{2n-1}(\beta_{TM} r_m)}{\beta_{TM}} \frac{\varepsilon_{||}}{\varepsilon_{\perp}}, \quad \mathbf{J}'_{TE} = \frac{J'_{2n-1}(\beta_{TE} r_m)}{\beta_{TE}}$$

$$\mathbf{K} = \frac{nK_n(q r_m)}{q^2 r_m}, \quad \mathbf{K}' = \frac{K'_n(q r_m)}{q}$$

$$R = \begin{cases} \sin\theta_m, & \theta < \theta_c \\ -\cos\theta_m, & \theta > \theta_c \end{cases}, \quad T = \begin{cases} \cos\theta_m, & \theta < \theta_c \\ \sin\theta_m, & \theta > \theta_c \end{cases}$$

$$r_m = \begin{cases} \frac{a}{2\cos\theta_m}, & \theta < \theta_c \\ \frac{b}{2\sin\theta_m}, & \theta > \theta_c \end{cases}$$

$$\theta_m = \frac{m - \frac{1}{2}}{2N}\pi, \quad \theta_c = \arctan\frac{b}{a}$$

5.4 COMPARISON OF MODIFIED MARCATILI'S AND GOELL'S METHODS WITH EXPERIMENTAL RESULTS

We have carried out experiments in order to verify the calculated numerical results. Sapphire was selected as a material for dielectric rod waveguides, because it is a typical example of a uniaxial anisotropic dielectric. It is a nonmagnetic dielectric, therefore its μ is equal to that of vacuum. S-parameter measurements of a Sapphire dielectric waveguide have been carried out with vector network analyser HP 8510, using a direct connection of the input and output ports as a reference. A monocrystal Sapphire dielectric waveguide, oriented along the optical axis, with a cross section of 1.0×0.5 mm^2, with the total length of 112 mm and the tapering section length of 6 mm has been used for the vector measurements [14]. Experiments and simulations have shown that the best matching of an asymmetrically tapered dielectric waveguide and a metal waveguide is achieved when the tip is located on the axis of the metal waveguide (see more details in Chapter 7, where different types of transitions are discussed). The experimental setup is shown in Figure 5.7.

Experimental and numerical results of the Marcatili and Goell methods for a Sapphire waveguide are summarized in Figures 5.8 and 5.9.

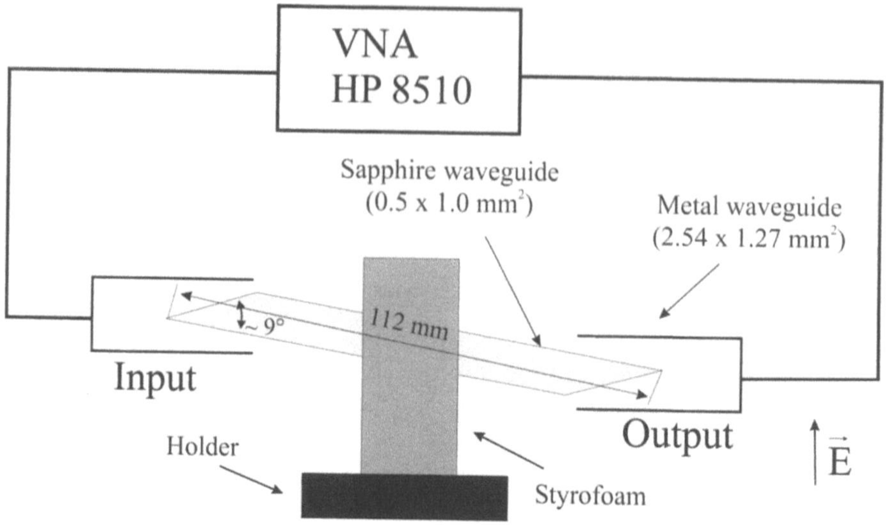

Figure 5.7. Schematic setup of vector measurements of dielectric waveguide sections.

Equations (5.36)–(5.38) and (5.52) have been used to obtain the propagation constant for uniaxial anisotropic waveguides for four combinations of $\varepsilon_{r\perp}$ and $\varepsilon_{r||}$: 11.56 and 11.56, 11.56 and 9.39, 9.39 and 11.56 (which corresponds to Sapphire [15]), and 9.39 and 9.39, respectively.

The comparison of modified Marcatili's and Goell's methods and with the experimental results are shown in Figure 5.10. One can see that the modified Goell method (Section 5.3.2, [12]) gives a better agreement with the experimental data in a wider frequency range, than the "modified Marcatili's method" (Section 5.3.1, [8]), described earlier.

The wavelength was measured directly at the frequency of 75 GHz by using a movable discontinuity (a rectangular metal "ring") in order to obtain a reference point. The phase values were corrected ($-2\pi n$ correction) to obtain a continuous dependence of phase versus frequency. Assuming a constant dielectric waveguide length L, one can write:

$$Dk = \frac{\Delta\phi}{L}, \tag{5.53}$$

where $\Delta\phi$ is the phase shift change when the frequency changes by a small step to the next point, and it is a change in the propagation constant in the dielectric waveguide. After obtaining the wavelength at one point by using the phase data, we calculate the propagation constants

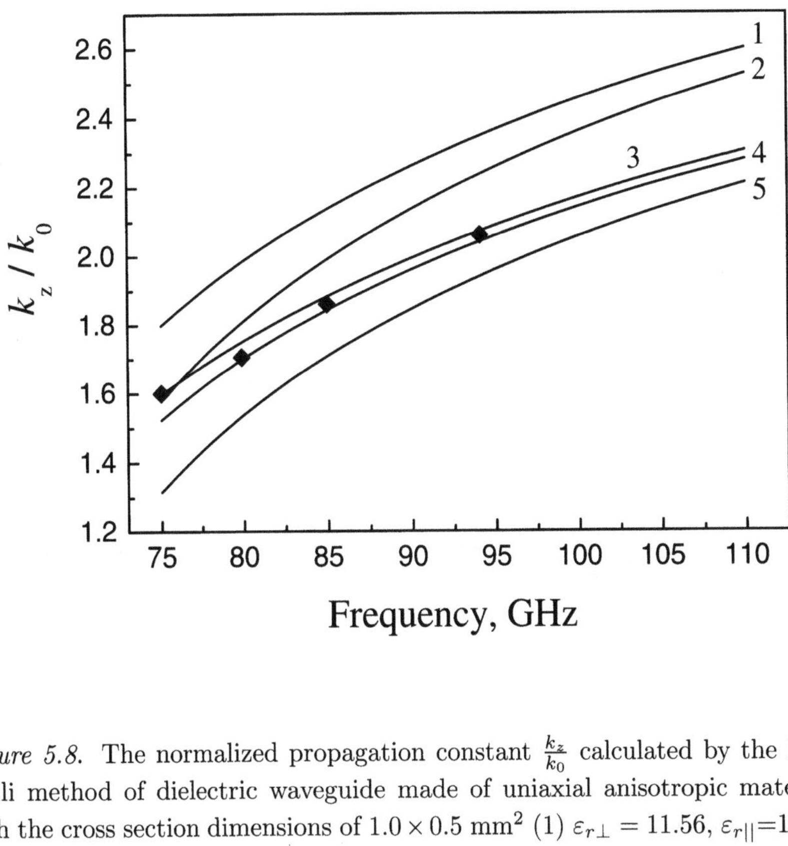

Figure 5.8. The normalized propagation constant $\frac{k_z}{k_0}$ calculated by the Marcatili method of dielectric waveguide made of uniaxial anisotropic materials with the cross section dimensions of 1.0×0.5 mm^2 (1) $\varepsilon_{r\perp} = 11.56$, $\varepsilon_{r||}$=11.56; (2) $\varepsilon_{r\perp} = 11.56$, $\varepsilon_{r||} = 9.39$; (3) calculated based on the experimentally measured phase results; (4) $\varepsilon_{r\perp} = 9.39$, $\varepsilon_{r||} = 11.56$; (5) $\varepsilon_{r\perp} = 9.39$, $\varepsilon_{r||} = 9.39$; ♦ — directly measured experimental points.

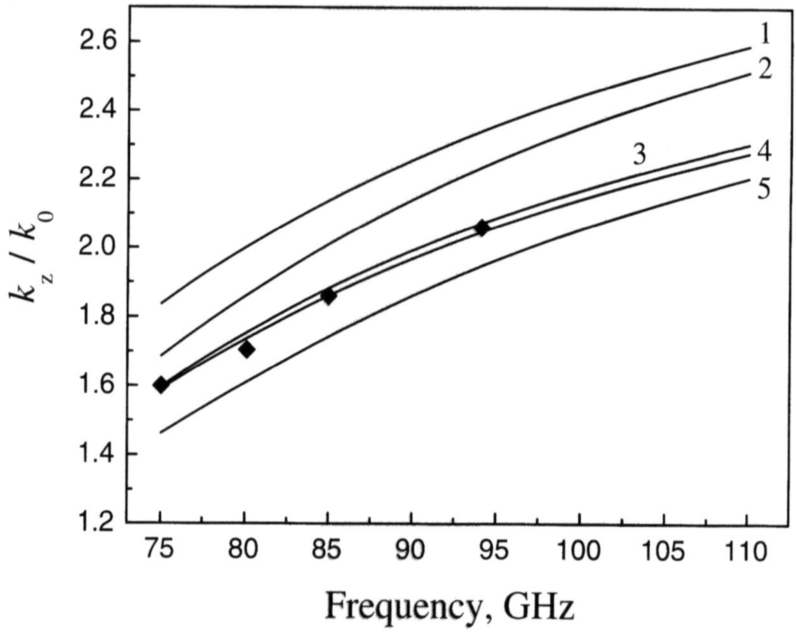

Figure 5.9. The normalized propagation constant $\frac{k_z}{k_0}$ calculated by Goell method of rectangular dielectric waveguides made of uniaxial anisotropic materials with the cross section dimensions of 1.0×0.5 mm^2 (1) $\varepsilon_{r\perp} = 11.56$, $\varepsilon_{r||}$=11.56; (2) $\varepsilon_{r\perp} = 11.56$, $\varepsilon_{r||} = 9.39$; (3) calculated based on the experimentally measured phase results; (4) $\varepsilon_{r\perp} = 9.39$, $\varepsilon_{r||} = 11.56$; (5) $\varepsilon_{r\perp} = 9.39$, $\varepsilon_{r||} = 9.39$; ♦ — directly measured experimental points.

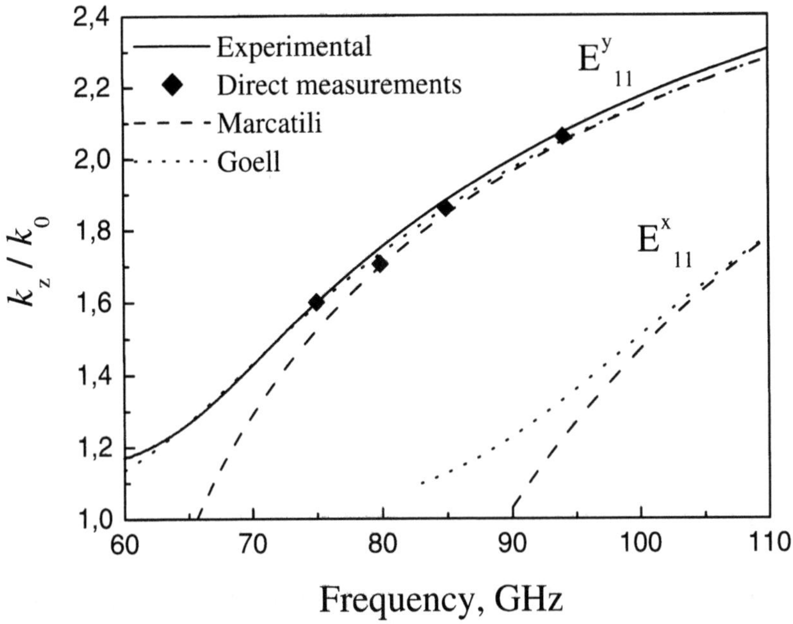

Figure 5.10. Comparison of the normalized propagation factor $\frac{k_z}{k_0}$ calculated by modified Marcatili's and Goell's methods and experimental results for a Sapphire dielectric waveguide with the cross section dimensions of 1.0×0.5 mm^2. Also, calculated data for the orthogonal x-polarization are shown.

at other frequencies (Figures 5.8 and 5.9, curves 3). For comparison, the normalized propagation constant measurements were repeated at 80, 85, 90, and 94 GHz. One can see in Figure 5.8 (curves 3 and 4) that the modified Marcatili method for the Sapphire dielectric rod waveguide gives a good agreement with the experimental data. The fact that the theoretical curve lies below the experimental one can be explained by the approximate nature of Marcatili's method. Comparing curve 1 with 2, and 4 with 5, one can see that the anisotropy changes the propagation characteristic considerably. The dispersion is increased, as with curve 2, when $\varepsilon_{||}$ is smaller than $\varepsilon_{\perp}$, or decreased when $\varepsilon_{||}$ is larger than $\varepsilon_{\perp}$, as with curve 4. The latter could be explained as follows. When the frequency is very high, there is almost no longitudinal electric field component, and the propagation constant is determined mainly by $\varepsilon_{\perp}$.

When the frequency drops, the longitudinal component of the electric field becomes larger, therefore the effect of larger $\varepsilon_{||}$ becomes stronger and thus the normalized propagation factor $\frac{k_z}{k_0}$ increases. Similarly, one can explain why the dispersion seen in the case of curve 2 is stronger.

Realizations of the modified Marcatili and Goell methods as Matlab codes for rectangular open dielectric waveguides with uniaxial cores (the axis is parallel to the waveguide axis) are provided on disk.

References

[1] A.Sv. Sudbø, Why are accurate computations of mode fields on rectangular dielectric waveguides difficult?, *J. of Lightwave Technology*, vol. 10, no. 4, April 1992, pp. 418-419.

[2] T. Itoh, *Numerical Techniques for Microwave and Millimeter-Wave Passive Structures*, New York: Wiley, 1989.

[3] E.A.J. Marcatili, Dielectric rectangular waveguide and directional coupler for integrated optics, *Bell System Technical J.*, vol. 48, 1969, pp. 2071-2102

[4] D. Marcuse, *Theory of Dielectric Optical Waveguides*, New York: Academic Press, 1974.

[5] J.E. Goell, A circular-harmonic computer analysis of rectangular dielectric waveguides, *Bell System Technical J.*, September 1969, pp. 2133-2160.

[6] G.I. Veselov and G.G. Voronina, To the calculation of open dielectric waveguide with rectangular cross-section, *Radiofizika*, vol. XIV, no. 12, 1971, pp. 1891-1901 (in Russian).

[7] S. Garcia, T. Hung-Bao, R. Martin, and B. Olmedo, On the application of finite methods in time domain to anisotropic dielectric waveguides, *IEEE Transactions on Microwave Theory and Techniques*, vol. 44, December 1996, pp. 2195-2206.

[8] S.N. Dudorov, D.V. Lioubtchenko, and A.V. Räisänen, Modification of Marcatili's method for the calculation of anisotropic rectangular dielectric waveguides, *IEEE Transactions on Microwave Theory and Techniques*, vol. 50, no. 6, 2002, pp. 1640-1642.

[9] R.M. Knox and P.P. Toulios, Integrated circuits for the millimeter through optical frequency range, *Proc. of Symposium on Submillimeter Waves,* 1970, Polytechnic Institute of Brooklyn, New York, pp. 497-516.

[10] M.I. Oksanen, S.A. Tretyakov, and I.V. Lindell, Vector circuit theory for isotropic and chiral slabs, *J. of Electromagnetic Waves and Applications*, vol. 4, 1990, pp. 613-643.

[11] T. Itoh, Dielectric waveguide-type millimeter-wave integrated circuits, in *Infrared and Millimeter Waves,* vol. 4, chapter 5, New York: Academic Press, 1981, pp. 199-273.

[12] S.N. Dudorov, D.V. Lioubtchenko, J.A. Mallat, and A.V. Räisänen, Modified Goell's method for the calculation of uniaxial anisotropic rectangular dielectric waveguides, *Microwave and Optical Technology Letters*, vol. 32, no. 5, 2002, pp. 373-376.

[13] Y. Kobayashi and T. Tomohiro, Resonant modes in shielded uniaxial-anisotropic dielectric rod resonators, *IEEE Transactions on Microwave Theory and Techniques*, vol. 41, 1993, pp. 2198-2205.

[14] D.V. Lioubtchenko, S. Dudorov, J. Mallat, J. Tuovinen, and A.V. Räisänen, Low loss sapphire waveguides for 75–110 GHz frequency range, *IEEE Microwave and Wireless Components Letters,* vol. 11, no. 6, 2001, pp. 252-254.

[15] V.V. Parshin, Dielectric materials for gyrotron output windows, *International J. of Infrared and Millimeter Waves*, vol. 15, 1994, pp. 339-348.

Chapter 6

Fabrication and measurements

6.1 METHODS FOR MATERIAL TESTING

The National Physical Laboratory (UK) has carried out a complex experiment on intercomparison of measurement techniques of 13 research groups from different countries using the same specimens in the frequency range of $30 - 900$ GHz ($\lambda = 10 - 0.33$ mm). The compared measurement methods were the dispersive Fourier transform spectroscopy, open Fabry-Perot resonators, the optically pumped laser spectroscopy, free space transmission and reflection measurements, and four-port and six-port reflectometers. The disagreement in the measured results has been up to 20 per cents for the real part of the permittivity ε and an order of magnitude for the loss factor $\tan\delta$.

There are two different approaches to the problem of measurements of the permittivity and the loss tangent or the refractive index and the absorption coefficient of dielectrics. One possibility is to use a noise or a quasi-noise signal source. Different variations of the Fourier spectroscopy use such signals. The other possibility is using a monochromatic signal source. The measurement methods with monochromatic sources can be further classified as "resonant" and "non-resonant" methods. These two approaches are widely used in the investigation of dielectrics from DC up to the optical frequency region and even at higher frequencies. However, every frequency range dictates its own measurement methods and specific realizations of measuring systems.

Non-resonant methods for measuring the refractive index and the loss tangent in the microwave range when applied at higher frequencies transform into the classical measuring method on the base of quasi-optical beams using different variants of Michelson interferometers (Figure 6.1). In the most common setup the dielectric plate under test is situated between transmitting and receiving antennas and the measured parameters are the complex reflection or/and transmission coefficients.

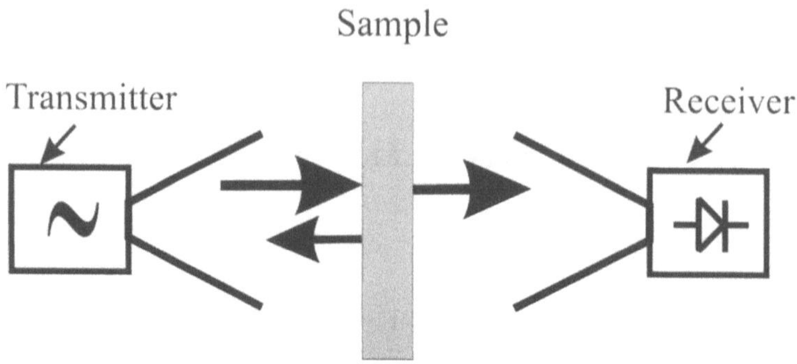

Figure 6.1. Non-resonance setup for refractive index and loss tangent measurements.

The main advantages of this method are the following: a wide frequency range, relative simplicity of realization, and a possibility for automatization of the measurement process. But the accuracy of the refractive index measurements is not very high: It is about 10^{-3}, because the information about the material parameters is extracted from a single reflection or/and transmission of the wave and the slab is normally thin. The accuracy of absorption (loss tangent) measurement is strongly dependent on the absorption level in the sample, and it is usually of the order of a few percent for the values of the loss tangent in the range $10^{-1} - 10^{-3}$. Measurements of absorption in high quality dielectrics with the loss tangent about 10^{-4} and less using this method are practically impossible, especially in thin plates.

6.2 OPEN FABRI-PEROT RESONATORS FOR MATERIAL TESTING IN THE MILLIMETER-WAVE REGION

The resonant methods of measuments have a better accuracy and sensitivity, but they are more complicated to implement. They are based on the use of different types of resonators, and the measured parameters are the resonant frequency and the quality factor of the loaded and unloaded resonator. The millimeter-wave range dictates the type of resonators that can be used. It is possible to use only quasi-optical open resonators with a large number of eigenmodes. The reason for this is that the sizes of closed cavity resonators are too small, and the quality factors are too low because of high losses in metal walls. The main type of resonators used in the millimeter-wave range is the Fabry-Perot resonator. Perhaps it is the simplest variant, but this is a very effective solution. It contains only two mirrors (Figure 6.2) and a coupling system. The quality factor of an open resonator with the distance between the mirrors about 300 mm can be up to 600 000 [3], so the accuracy and sensitivity of installations which are based on this type of resonator are very good.

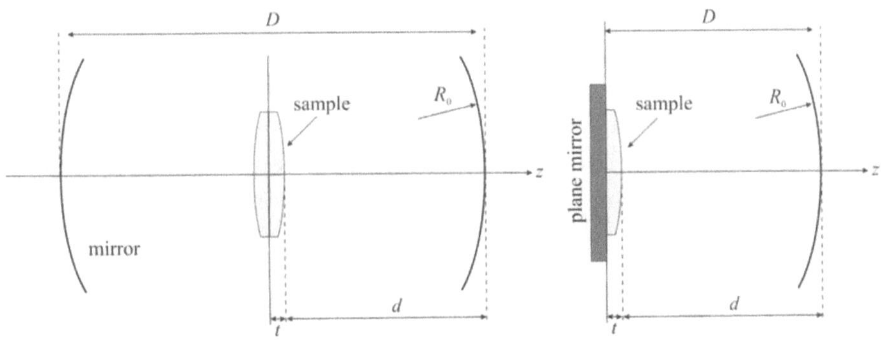

Figure 6.2. Schematic drawing of spherical and semispherical resonators.

Two main types of installations are used to determine the refractive index. One of them is a resonator with the fixed length fed by a frequency-scanning signal source [1]. In this case the measured parameters are the resonant frequency and the quality factor, determined through the resonance curve width. The other possibility is to use a signal source with a fixed frequency and a resonator with a variable length. In this case the measured parameter is the change of the resonator length [2]. Combinations of these two approaches are possible. In both cases it is necessary to know the thickness of the material sample (usually it is a planar slab). The resonator length and the radius of the mirror curvature are determined from separate calibration measurements. The information about the absorption in dielectric ($\tan\delta$) is extracted from the difference of the quality factor of the empty and loaded resonators, knowing the sample thickness. Furthermore, there is a rather modern method for dielectric parameters measurements in wide frequency ranges, practically in millimeter and submillimeter frequencies, which is free from many disadvantage mentioned above [3]. The method is based on the determination of the resonant frequency of a resonator with a plain wafer placed normally to its axis at the waist plane of resonator. At a resonance, the plate optical thickness in this case is a multiple of half-wavelength. Having measured the resonant frequency and determined the number of half-wavelengths one can calculate the refractive index and, separately, the plate thickness. The method of finding this frequency is rather efficient and, in fact, is the basis of the method of measurements.

6.2.1 Classical theory and its extensions

The fundamental theory of spherical open resonators can be found e.g. in [5]. In a sense, a semispherical open resonator is similar to a spherical resonator with the double length and antisymmetrical modes only. For the situation when a dielectric sample is placed into the resonator, the fundamental theory is well established in [6]. The resonant frequency of an empty resonator can be found from the following formula [5]:

$$\frac{f}{f_0} = (q+1) + \frac{2p+l+1}{\pi} \arccos\left(1 - \frac{D}{R_0}\right) \tag{6.1}$$

where $f_0 = \frac{c}{2D}$, c is the speed of light, R_0 is the curvature radius of the mirrors, D is the distance between them, q is the number of halves of the wavelength of the standing wave in the resonator, p and l are the higher-order mode numbers (Figure 6.2).

In practice, higher-order modes are avoided, therefore p and l are zeros. When a dielectric sample is placed into the resonator, the following relations hold:

$$\frac{1}{n} \cot\left(nkt - \Phi_t\right) = \tan(kd - \Phi_d) \tag{6.2}$$

for symmetric modes and

$$\frac{1}{n} \tan\left(nkt - \Phi_t\right) = -\tan(kd - \Phi_d) \tag{6.3}$$

for antisymmetric modes, where n is the refractive index of the sample, t is the half of the sample thickness, Φ_t and Φ_d are phase correction coefficients:

$$\Phi_t = \arctan\left(\frac{2t}{nk\omega_0^2}\right) \tag{6.4}$$

$$\Phi_d = \arctan\left\{\frac{2}{k\omega_0^2}\left(d + \frac{t}{n^2}\right)\right\} - \arctan\left(\frac{2t}{n^2 k\omega_0^2}\right), \tag{6.5}$$

$$k\omega_0^2 = 2\sqrt{\left(d + \frac{t}{n^2}\right)\left(R_0 - d - \frac{t}{n^2}\right)}$$

where ω_0 is the gaussian beam radius. The definitions of d, D, and R_0 are shown in Figure 6.2. The above expressions are valid for both spherical and semispherical resonators.

As to the loss tangent, the following formula is used:

$$\tan\delta = \frac{1}{Q_\varepsilon}\frac{\Delta t + d}{\Delta t + \frac{1}{2k}\sin 2\left(kd - \Phi_d\right)} \tag{6.6}$$

where $1 \le \Delta \le \varepsilon_r$ and

$$\Delta = \frac{n^2}{n^2\sin^2(nkt - \Phi_t) + \cos^2(nkt - \Phi_t)}$$

for symmetric modes and

$$\Delta = \frac{n^2}{n^2\cos^2(nkt - \Phi_t) + \sin^2(nkt - \Phi_t)}$$

for antisymmetric ones. As the empty resonator has a finite quality factor Q_0, the quality factor of the resonator with the sample Q_L has to be corrected according to

$$\frac{1}{Q_\varepsilon} = \frac{1}{Q_L} - \frac{1}{Q_0} \tag{6.7}$$

An analysis of the influence of offsets of the sample, its tilt, and determination of its thickness is given in [10], and it shows that the errors due to these factors are minimized when the sample thickness is close to an integer number times half of the wavelength. This follows from the fact that in that case the electric field at the dielectric sample surfaces is almost zero.

The theory above is derived from an approximation, that the dielectric sample profile repeats the phase front of the gaussian beam, and after that frequency corrections are introduced, calculated using the perturbation theory:

$$\frac{\Delta f}{f} = -\frac{n^2 - 1}{n^2}\frac{t\sin^2(kd - \Phi_d)}{n^2k^2\omega_0^2(t\Delta + d)} \tag{6.8}$$

or, alternatively,

$$\frac{\Delta f}{f} = -\frac{n^2 - 1}{n^4}\frac{t\Delta\sin^2(nkt - \Phi_t)}{k^2\omega_0^2(t\Delta + d)} \tag{6.9}$$

which is more convenient. Instead of correcting the frequency one can alternatively introduce an effective sample thickness change [7]:

$$\delta t = \frac{\omega_t^2}{4R_t} \tag{6.10}$$

where $R_t = t + \frac{n^2(k\omega_0^2/2)^2}{t}$ is the curvature radius of the gaussian beam at the dielectric surface, $\omega_t^2 = \omega_0^2 \left[1 + \left(\frac{2t}{k\omega_0^2}\right)^2\right]$ is the gaussian beam radius at the same surface.

The above theory was intuitively modified within the optical approximation in [8]. The explanation was that for small incidence angles the gaussian bean curvature within the dielectric is smaller by a factor of n, therefore the phase constants have to be changed, namely, in (6.4) and (6.5) n has to be written instead of n^2. Later in [7], the same authors (see [6]) improved their theory, and the phase corrections became as follows:

$$\Phi_t = \arctan\left(\frac{t}{nz_0}\right) - \arctan\left(\frac{1}{kR_1(t)}\right) \tag{6.11}$$

$$\Phi_d = \arctan\left(\frac{d'}{z_0}\right) - \arctan\left(\frac{1}{kR_0}\right) - \arctan\left(\frac{t}{nz_0}\right) + \arctan\left\{\frac{1}{kR_2(t)}\right\} \tag{6.12}$$

where

$$R_1(t) = t + \frac{n^2 z_0^2}{t}$$

$$R_2(t) = \frac{t}{n} + \frac{nz_0^2}{t}$$

$$z_0^2 = \sqrt{d'(R_0 - d')}$$

$$d' = d + \frac{t}{n}$$

For the empty resonator this theory gives

$$f = \frac{c}{2D}\left[q + 1 + \frac{1}{\pi}\arccos\left(1 - \frac{D}{R_0}\right) - \frac{1}{2\pi k R_0}\right] \tag{6.13}$$

The last term is absent in the older theory. However, in practice such changes are rather small and usually the older theory is used for calculations.

The resonator theory is successfully employed e.g. in [9, 10]. The work [10] is especially useful for studying purposes, as it shows how to use the measured frequency-amplitude characteristics. The received power can be expressed as a Lorentz curve

$$|S_{21}(f)|^2 \approx \frac{|S_{21}|^2(f_r)}{1 + 4Q^2\left(\frac{f}{f_r} - 1\right)^2} \tag{6.14}$$

By introducing $y = \frac{1}{|S_{21}(f)|^2}$ one can rewrite this equation as

$$y = Af^2 + Bf + C \tag{6.15}$$

where

$$A = \frac{4Q^2}{|S_{21}(f_r)|^2 f_r^2}$$

$$B = -\frac{8Q^2}{|S_{21}(f_r)|^2 f_r}$$

$$C = \frac{1 + 4Q^2}{|S_{21}(f_r)|^2}$$

The resulting quadratic function is much easier to fit.

In [11] an improvement was proposed in order to decrease the errors in calculating refractive indices of samples by calculating from (6.13) the distance D and then calculating n from (6.2). In that case the errors tend to cancel each other.

Determination of the loss tangent, especially for low-loss materials, can be also improved as proposed in [12]. In that work the authors propose to move a dielectric sample, which is an integer number of halfs of the wavelength thick, along the resonator to place it so that the electrical field is at maximum at the dielectric surfaces. As a result, the absorbed power becomes n^2 times higher enabling measurements of very small loss tangents. For example, for a 3 mm thick sapphire plate, the loss tangent as low as $5.5 \cdot 10^{-7}$ is measurable with their installation.

Recent experimental results, carried out by these authors, show that 5 micron thin dielectric films on substrates are detectable, and a method to measure their properties is being developed and verified. In this method, the properties of thin films on dielectric substrates can be measured without a priori information about the substrate and film thicknesses.

6.3 MATERIALS FOR MILLIMETER-WAVE DIELECTRIC WAVEGUIDES

If a dielectric material has the loss tangent of about 10^{-4} and a relative permittivity of 10 – 15, the waveguide losses can be expected to be in the order of 5 dB/m at 150 GHz while 10 – 12 dB/m is typical for the standard metal waveguide at this frequency. The properties of some dielectric materials which appear to be appropriate for millimeter-wave dielectric waveguides are summarized in Table 6.1. There are some variations in data taken from different sources (see e.g., [4, 13, 14]), because all the parameters depend on the material growth conditions, the method of its doping, measurement method, etc.

Table 6.1. Dielectric properties of some materials at $\approx$ 140 GHz frequency

Materials	Permittivity, ε	$\tan\delta$, $\times 10^{-4}$
Si	11.6–11.8	0.01
Diamond	5.6644	0.08
GaAs	10.9–13.03	3
BN	3.1791	8–15
Quartz	4.44–4.64	0.9–1.8
Teflon	1.96–2	3–4
Al2O3	10.5	2.7–3.2
AlN	7.7–9.5	4.5
Ferrite	13.3(NiZn), 15.6(Li), 10.05(FeY)	1.24–7.5
Sapphire	9.39–11.56	1.1–2.1

New developments in the chemical vapor deposition technique allow producing Diamond disks [1] with the loss tangent about $10^{-5} - 10^{-6}$, which makes them very attractive. Also wide band gap semiconductors AlN and GaN are now popular and are being investigated by many scientists (e.g., [15]). These materials can have a very high resistivity and perhaps a very low loss tangent. However, there are no experimental data available in the literature for monocrystal samples of these materials because of difficulties in growing them.

The most important parameters are the loss tangents measured at frequencies 120–170 GHz. Looking at the data on Si in [16], it is possible to calculate that its resistivity 10^4 $\Omega\cdot$cm corresponds to $\tan\delta \approx 0.7\cdot 10^{-4}$ compared with $0.7\cdot 10^{-4}$ measured in [16]. In comparison, it can be seen that in spite of that GaAs has a higher DC resistivity than Si, its loss tangent is higher than that of Si [16]. This means that a high DC resistivity does not necessarily predict low losses at millimeter wavelengths. In [17] the properties of different dielectric materials can be found including GaAs and Si with DC resistivity ≈8 kΩ·cm. This resistivity of the silicon sample does not differ much from the resistivity of 11 kΩ·cm for the silicon sample in [18], but the values of the loss tangents are very different. However, Si may have a very low loss tangent. Si doped with boron with the loss tangent $4 \cdot 10^{-5}$ [18] can be appropriate. Moreover, an interesting effect was described in [19]: the dielectric loss behavior of silicon does not degrade after electron or neutron irradiation but is even improving. Another way to improve the dielectric properties of silicon is traditional; that is, to compensate the conductivity by doping with acceptors. In [19] gold was used as an acceptor and also the effect of electron and neutron irradiation was investigated. The loss tangent for all the Si samples is shown graphically, and it drops nearly inversely proportional to the frequency. Therefore, one can expect that the dielectric properties of such silicon samples will be even more attractive at frequencies above 145 GHz. Thus, one can conclude that materials with a relatively high dielectric constant and a low loss tangent at millimeter-wave frequencies exist. An extensive review on this subject can be found in [20].

References

[1] R. Heidinger, G. Dammertz, A. Meier, and M.K. Thumm, CVD diamond windows studied with low- and high-power millimeter waves, *IEEE Transsactions on Plasma Science*, vol. 30, no. 3, 2002, pp. 800-807.

[2] M.N. Afsar, H. Ding, and K. Tourshan, A new 60 GHz open-resonator technique for precision permittivity and loss-tangent measurenent, *IEEE Transactions on Instrumentation and Measurement*, vol. 48, no. 2, 1999, pp. 626-630.

[3] A.F. Krupnov, M.Yu. Tretyakov, V.V. Parshin, V.N. Shanin, and S.E. Myasnikova, Modern millimeter-wave resonator spectroscopy of broad lines, *J. Molecular Spectroscopy*, vol. 202, 2000, pp. 107-115.

[4] V.V. Parshin, Dielectric materials for gyrotron output windows, *International J. Infrared and Millimeter Waves*, vol. 15, no. 2, 1994, pp. 339-348.

[5] H. Kogelnik and M. Hill, Modes in optical resonators, in A.K. Levine (ed.), *Lasers*, vol. 2, chapter 5, New-York: Dekker, 1966.

[6] A.L. Cullen and P.K. Yu, The accurate measurement of permittivity by means of an open resonator, *Proc. of the Royal Society of London*, vol. A325, 1971, pp. 493-509.

[7] A.L. Cullen and P.K. Yu, Measurements of permittivity by means of an open resonator. I. Theoretical, *Proc. of the Royal Society of London*, vol. A380, 1982, pp. 49-71.

[8] A.C. Lynch, Open resonator for the measurements of permittivity, *Electronics Letters*, vol. 14, no. 18, 31 August, 1978, p. 596.

[9] B. Komiyama, M. Kiyokawa, and T. Matsui, Open resonator for precision measurements in the 100 GHz band, *IEEE Transactions on Microwave Theory and Techniques*, vol 39, no. 10, 1991, pp. 1792-1796.

[10] T.M. Hirvonen, P.Vainikainen, A. Lozowski, and A.V. Räisänen, Measurement of dielectrics at 100 GHz with an open resonator connected to a network analyzer, *IEEE Transactions on Instrumentation and Measurement*, vol. 45, no. 4, 1996, pp. 780-786.

[11] A.L. Cullen, P. Nagenthiram, and A.D. Williams, Improvement in open-resonator permittivity measurement, *Electronics Letters*, vol 8, no. 23, 16 November 1972, pp. 577-579.

[12] A.F. Krupnov, V.N. Markov, G.Y. Golubyatnikov, I.I. Leonov, Y.N. Konoplev, and V.V. Parshin, Ultra-low absorption measurement in dielectrics in millimeter- and submillimeter-wave range, *IEEE Transactions on Microwave Theory and Techniques*, vol. 47, no. 3, 1999, pp. 284-289.

[13] R.C. Weast (editor), *Handbook of Chemistry and Physics,* Cleveland, OH: CRC Press, 57th edition, 1976-1977.

[14] R.F. Davis, III-V nitrides for electronic and optoelectronic applications, *Proc. of the IEEE*, vol. 79, no. 5, 1991, pp. 702-712.

[15] M.N. Yoder, Gallium nitride past, present, and future, *Proc. of the Conference on Advanced Concepts in High Speed Semiconductor Devices and Circuits, IEEE/Cornell*, 1997, pp. 3-12.

[16] V.V. Meriakri, B.A. Murmuzhev, and M.P. Parkhomenko, Millimeter wave devices based on dielectric, ferrite and semiconductor waveguides, *Proc. of the Microwave and Optoelectronics Conference*, 1997, vol. 2, pp. 431-433.

[17] M.N. Afsar and K.J. Button, Millimeter-wave dielectric measurement of materials, *Proc. of the IEEE*, vol.73, no. 1, January 1985, pp. 131-153.

[18] M.N. Afsar, H. Chi, and X. Li, Millimeter wave complex refractive index, complex dielectric permittivity and loss tangent of high purity and compensated silicon, *Conference on Precision Electromagnetic Measurements*, 1990, CPEM'90 Digest, pp. 238-239.

[19] J. Molla, R. Vila, R. Heidinger, and A. Ibarra, Radiation effects on dielectric losses of Au-doped silicon, *J. of Nuclear Materials*, vol. 258-263, 1998, pp. 1884-1888.

[20] V.V. Meriakri, Low-loss materials for application in millimetre wave ranges, in *The Science and Technology of Millimetre Wave Components and Devices,* Ed. by V.E. Lyubchenko, London and New York: Taylor & Francis, 2002, pp. 117-132.

Chapter 7

Excitation of millimeter-wave dielectric waveguides: computer simulations and experiments

Excitation of dielectric rod waveguides (DRW) can be somewhat difficult and problematic, because in dielectric rod waveguides the electromagnetic field of propagating waves is partially outside the rod, as shown in Figure 5.1. Previously, dielectric waveguides were made of such materials as polyethylene and teflon, which have a low dielectric constant. Therefore, the dimensions of such waveguides could be chosen to coincide with the cross section dimensions of the standard metal waveguide. In order to decrease losses in transitions, the authors usually designed a taper section followed by a small horn, see e.g., [1] (Figure 7.1), so that a contact between the metal walls of the waveguide and the dielectric rod was not avoided but in fact was rather desirable for mechanical reasons.

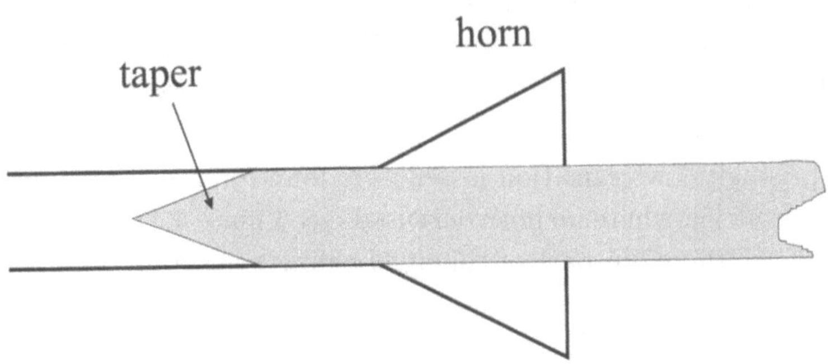

Figure 7.1. A transition from a metal to a dielectric waveguide with the same cross sections.

In the case of a dielectric waveguide made of, for example, a semiconductor or a dielectric with a high permittivity, the cross section dimensions of the dielectric rod required for the fundamental-only mode operation become smaller. Therefore, other constructions for dielectric waveguide mode launchers are used [2–4], see Figures 7.2 and 7.3.

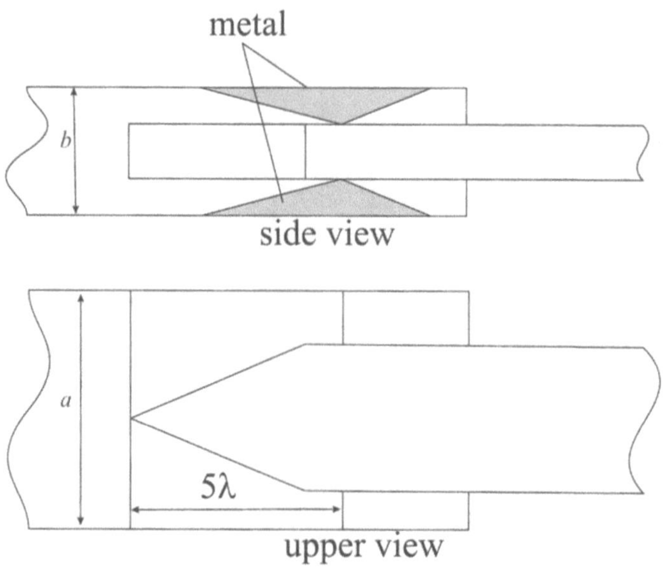

Figure 7.2. A transition from a metal to a silicon waveguide [1].

All such constructions consist of the following parts: a transition from the standard metal waveguide to the waveguide filled with a dielectric, followed by a transition to the dielectric waveguide itself. In the structure presented in Figure 7.3 [2], and also e.g. in [4, 5], there is neither a horn nor a contact between the metal waveguide and the dielectric rod waveguide.

Surprisingly low transition losses were found in a mode launcher for phase shifters [5], where no horn was used (see Figure 4.4). The dielectric waveguide as the main part of this phase shifter is simply a rectangular ferrite rod tapered at the ends and placed on a substrate. The area of the cross section of this rod is smaller than a quarter of that of the standard waveguide (1.3×2.6 mm^2 vs. 7.2×3.6 mm^2).

In this chapter we present the results of our simulations and experimental measurements of rectangular cross section dielectric rod waveguides made of such materials as monocrystal Sapphire, oriented along

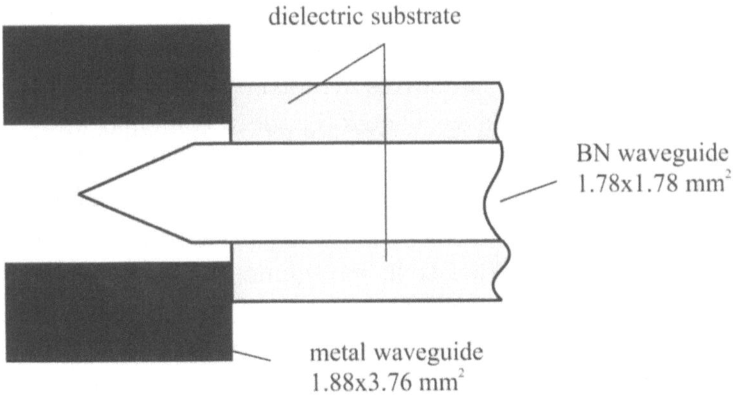

Figure 7.3. A transition from a metal to a boron nitride waveguide [2].

the optical axis, and GaAs.

7.1 COMPUTER SIMULATIONS WITH THE FINITE ELEMENT METHOD

7.1.1 Tapers of the dielectric waveguide

Three types of transitions (taper sections) from the standard metal waveguide to open dielectric rod waveguides with relatively high dielectric permittivities seem to be the most suitable (Figure 7.4). The "dove tail taper" [6] was not considered due to technical difficulties of manufacturing from fragile materials like Sapphire or Si.

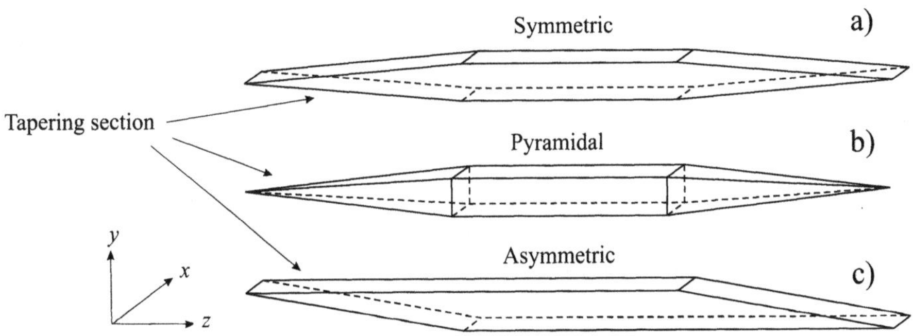

Figure 7.4. Three types of taper sections: a) symmetrical, b) pyramidal, c) asymmetrical.

Eigenwaves of two orthogonal polarizations can propagate in dielectric rod waveguides, which results in a possibility of transformations from one polarization to the other. One way to decrease this effect is to design the waveguide so that waves of different polarizations have different propagation constants. Therefore, the rectangular cross section is more preferable than the square one. The ratio $a/b = 0.5$ appears to be suitable (the geometry and the coordinate system is defined in Figure 5.4). Another problem is that if we increase the horizontal and/or vertical dimension of the dielectric waveguide, we can have multimode propagation, which is not desirable. It is well known that, in order to have a good transition matching, we should have the field distribution in the dielectric waveguide as close as possible to that in the metal waveguide. From this consideration it is preferable to orient the wide side of the dielectric waveguide parallel to the short side of the metal waveguide, as in this case the uniform field of the fundamental mode of the metal guide is better matched with the field distribution in the dielectric rod. Moreover, according to calculations, the mode E_{11}^{y} has a larger propagation constant (i.e., a stronger field concentration) than that of the E_{11}^{x} mode. Therefore, the vertical dimension of the dielectric waveguide is chosen to be larger and the wider wall to be perpendicular to that of the metal waveguide.

The rectangular cross section for the frequency range of 75 – 110 GHz was chosen initially according to the recommendations in [4] ($k_0 b = 1.7 - 1.9$, $a/b = 0.5$).

The Finite Element Method (FEM) implemented in *Agilent HFSS* was used to simulate the S-parameters of different configurations of transitions to rectangular dielectric rod waveguides with a high permittivity and a low loss tangent. The structure, as simulated with HFSS, is shown in Figure 7.5. The standard WR-10 (2.54×1.27 mm^2) metal waveguides were used as the input and output circuits. The frequency range was 75 – 110 GHz (W-band). The material used in this simulation was isotropic and had the refractive index $n = 3$. Due to the symmetry of the structure, it is possible to consider only one quarter of the whole configuration and save the computing resources by adding two symmetry planes (in terms of the HFSS program).

E and H tapering planes

Both symmetrical and asymmetrical tapers in the E- and H-planes have been simulated (Figure 7.6) in order to determine which type of tapers (Figure 7.4) is the most preferable from the viewpoint of transmission

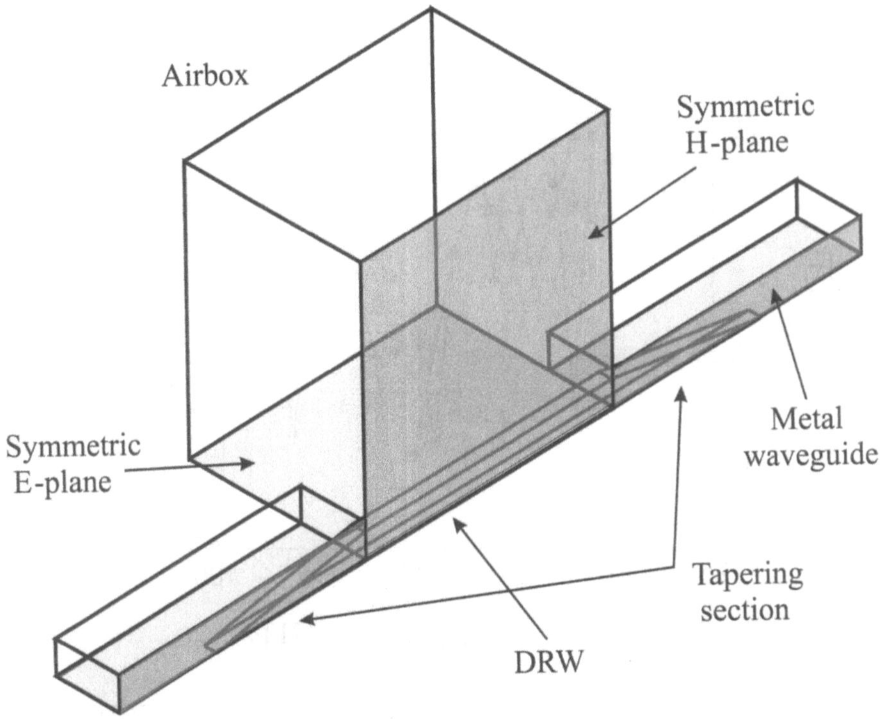

Figure 7.5. HFSS simulation structure.

characteristics. One can see in Figure 7.6 that there is no significant difference between the results for symmetrical and asymmetrical tapers in the H-plane, and that there is a small difference in transmission characteristics between the results for tapers in H- and E-planes. However, the reflections for H-tapers are higher. Also one can see that the transmission coefficient at the lower end of the frequency range is rather low, possibly because the dielectric waveguide field spreads out too much.

To improve the S_{21} parameter we increase a little the cross section dimensions of the dielectric waveguide, to 0.55×1.1 mm^2 (Figure 7.7). The difference in the transmission characteristics for E- and H-plane tapers becomes more significant. Therefore, the E-plane taper section is more preferable, and in the following only E-plane taper sections will be simulated.

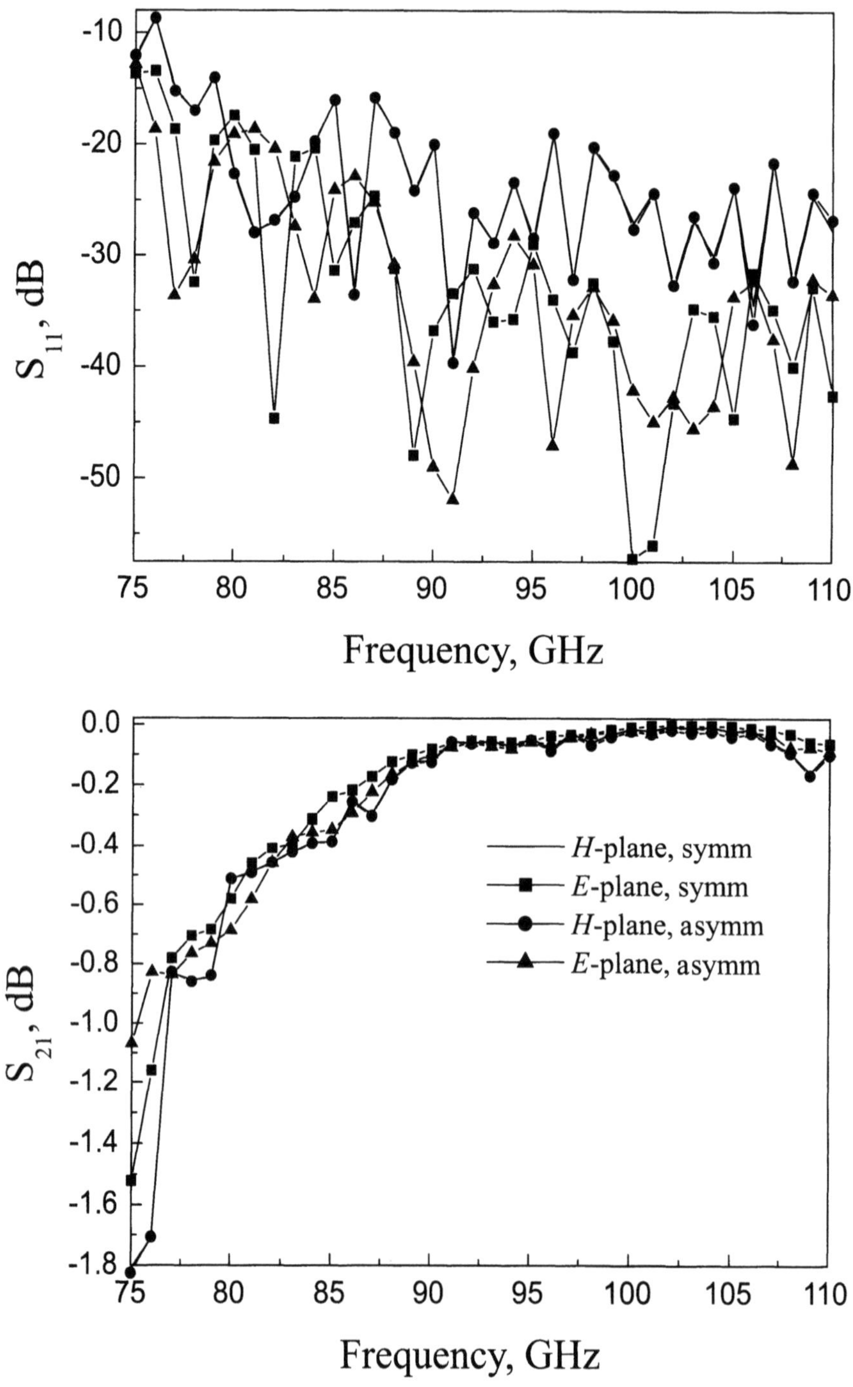

Figure 7.6. Simulation results for tapers in E- and H-planes. The dimensions of the dielectric rod waveguide are 1.0×0.5 mm^2.

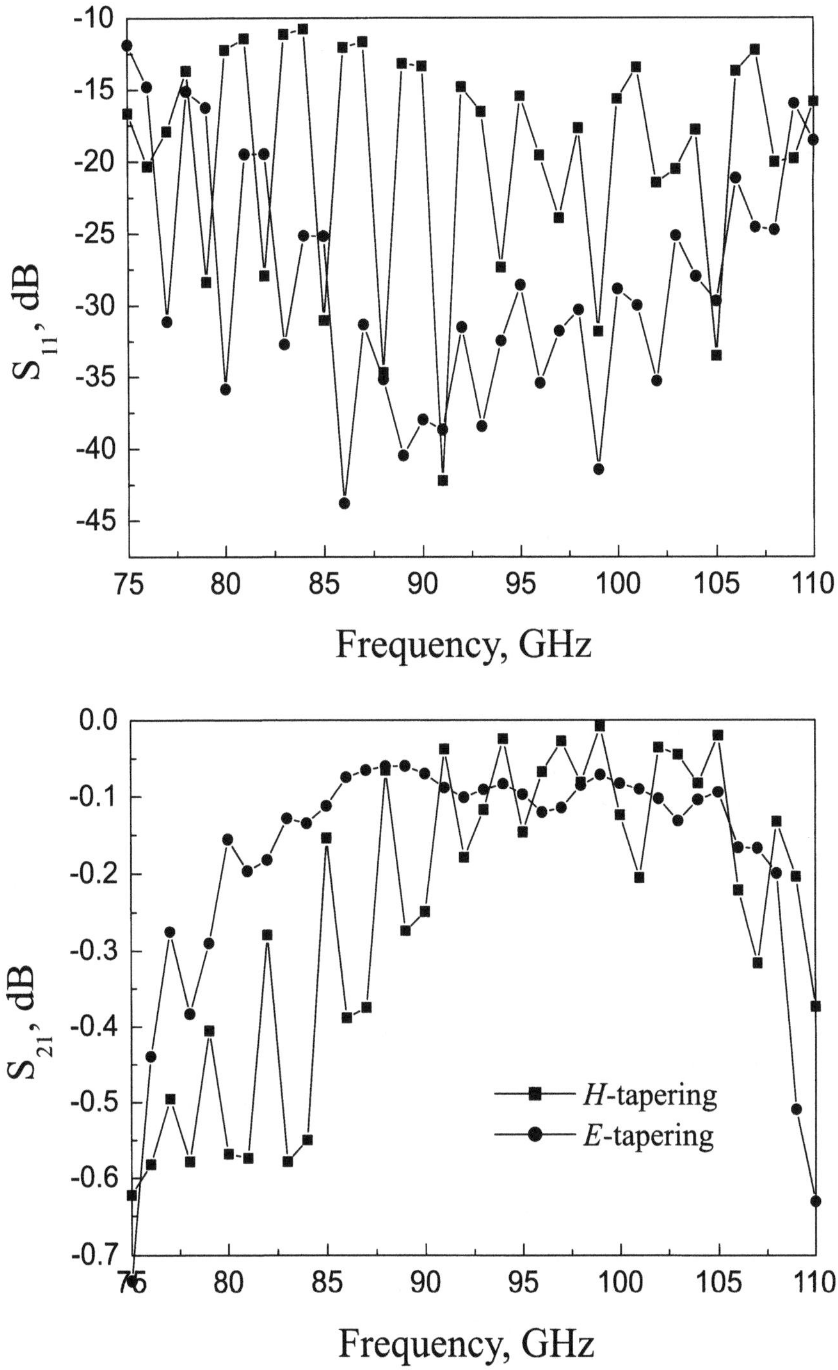

Figure 7.7. Simulation results for tapers in E- and H-planes. The dimensions of the dielectric rod waveguide are $1.1 \times 0.55\ \text{mm}^2$.

Pyramidal tapers

Tapers of the pyramidal type with different edge widths of the tip d were simulated in HFSS (Figure 7.8). One can see in Figure 7.8 that there is no significant difference between the results for the pyramidal type and for the symmetrical type of tapers, but technologically the symmetric taper is preferable, especially if the dielectric waveguide is made of a fragile material.

Different cross-sections of the dielectric waveguide

The simulation results for dielectric rod waveguides with the same symmetrical 6 mm taper but with the cross section sizes of 0.9×0.45 mm^2, 0.95×0.475 mm^2, 1.0×0.5 mm^2, 1.05×0.525 mm^2, 1.1×0.55 mm^2, 1.15×0.575 mm^2, and 1.2×0.6 mm^2 are shown in Figure 7.9. We can observe that an increase of the cross section size improves the characteristics at the lower frequency end, but if the size increases too much, this results in undesirable dips in the S_{21} characteristic, possibly due to excitation of higher-order modes. Moreover, the central frequency of the dips shifts to lower frequencies with increasing the cross section. The cross sections of 1.0×0.5 mm^2, 1.05×0.525 mm^2, and 1.1×0.55 mm^2 produce the most suitable S_{21} and S_{11} characteristics and further only 1.0×0.5 mm^2 and 1.1×0.55 mm^2 cross sections will be investigated.

Different taper lengths

The simulation results for dielectric rod waveguides with the dimensions 1.0×0.5 mm^2 and 1.1×0.55 mm^2 and a symmetrical E-taper with different taper lengths of $1 - 8$ mm, simulated in HFSS, are presented in Figure 7.10. One can see, that the tapers with the lengths smaller than 2 mm have relatively high reflection and transition losses, but the 2 mm taper is already acceptable. It even gives a better transmission coefficient at lower frequencies with the maximum approximately at 88 GHz. Increasing the taper length only slightly improves the transmission at higher frequencies, but at large lengths this improvement is very small. Moreover, at 8 mm an undesirable dip appears. Technically a long taper section is more difficult if the material is fragile, and it turns out to be unnecessary.

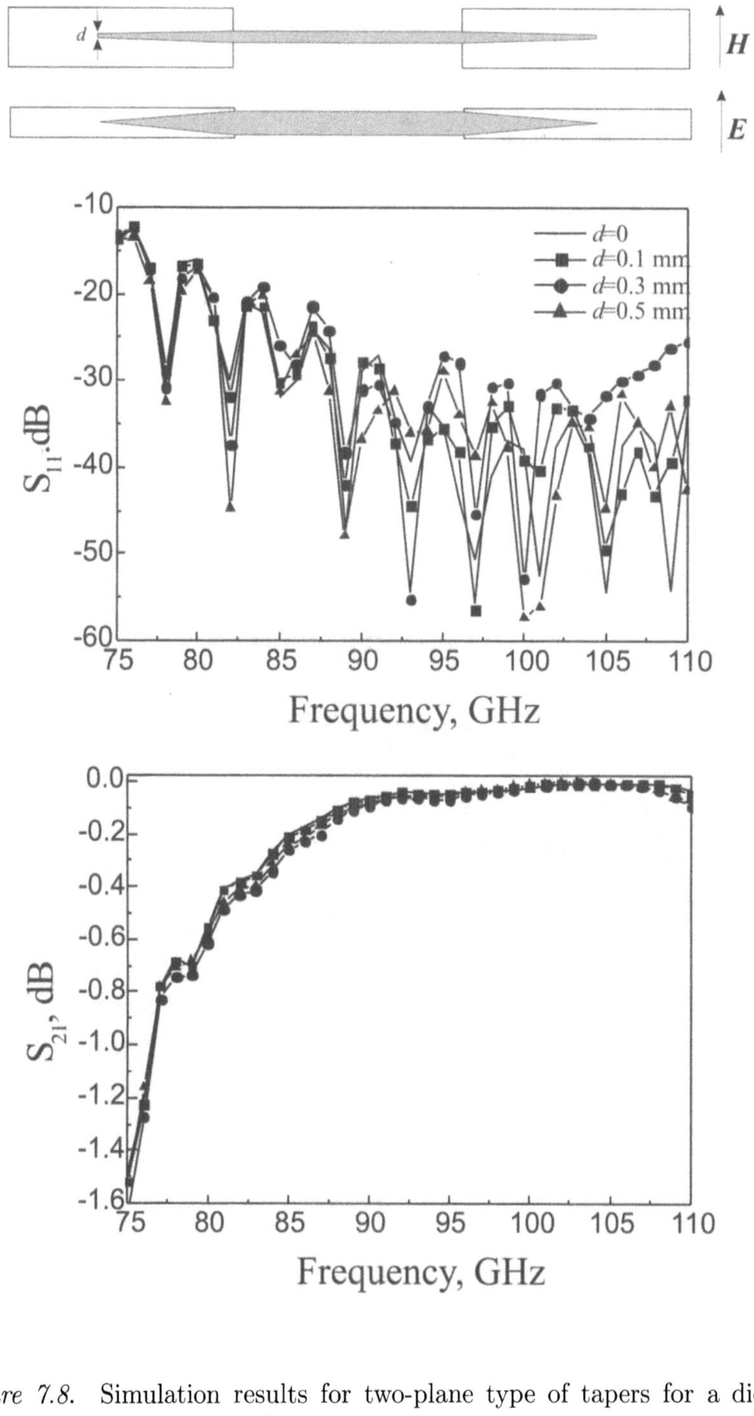

Figure 7.8. Simulation results for two-plane type of tapers for a dielectric waveguide with 1.0×0.5 mm^2 cross section. $d = 0$ corresponds to the pyramidal taper, and $d = 0.5$ mm corresponds to the E-taper ($l = 6$ mm).

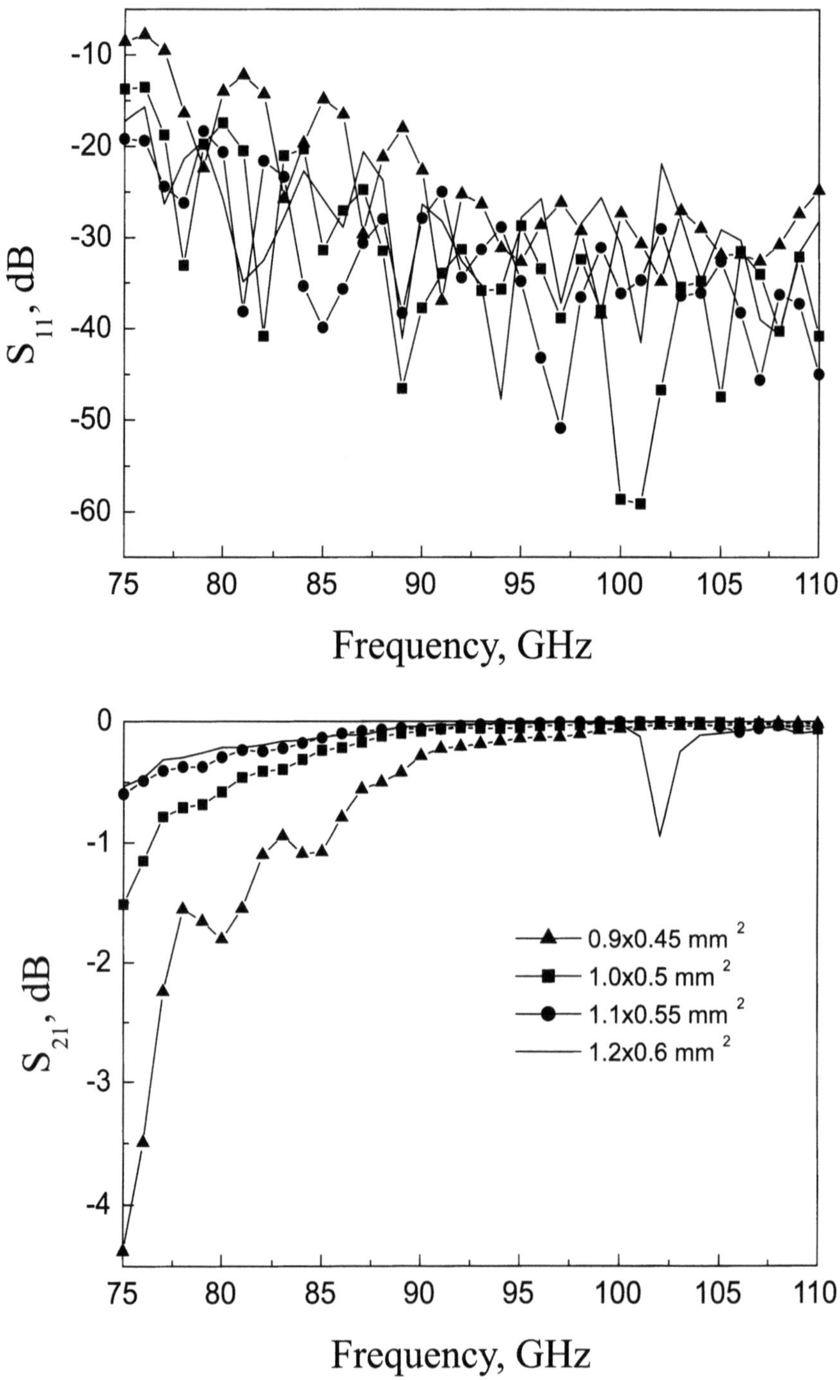

Figure 7.9. Simulation results for different cross sections of a dielectric waveguide with a 6 mm symmetrical taper.

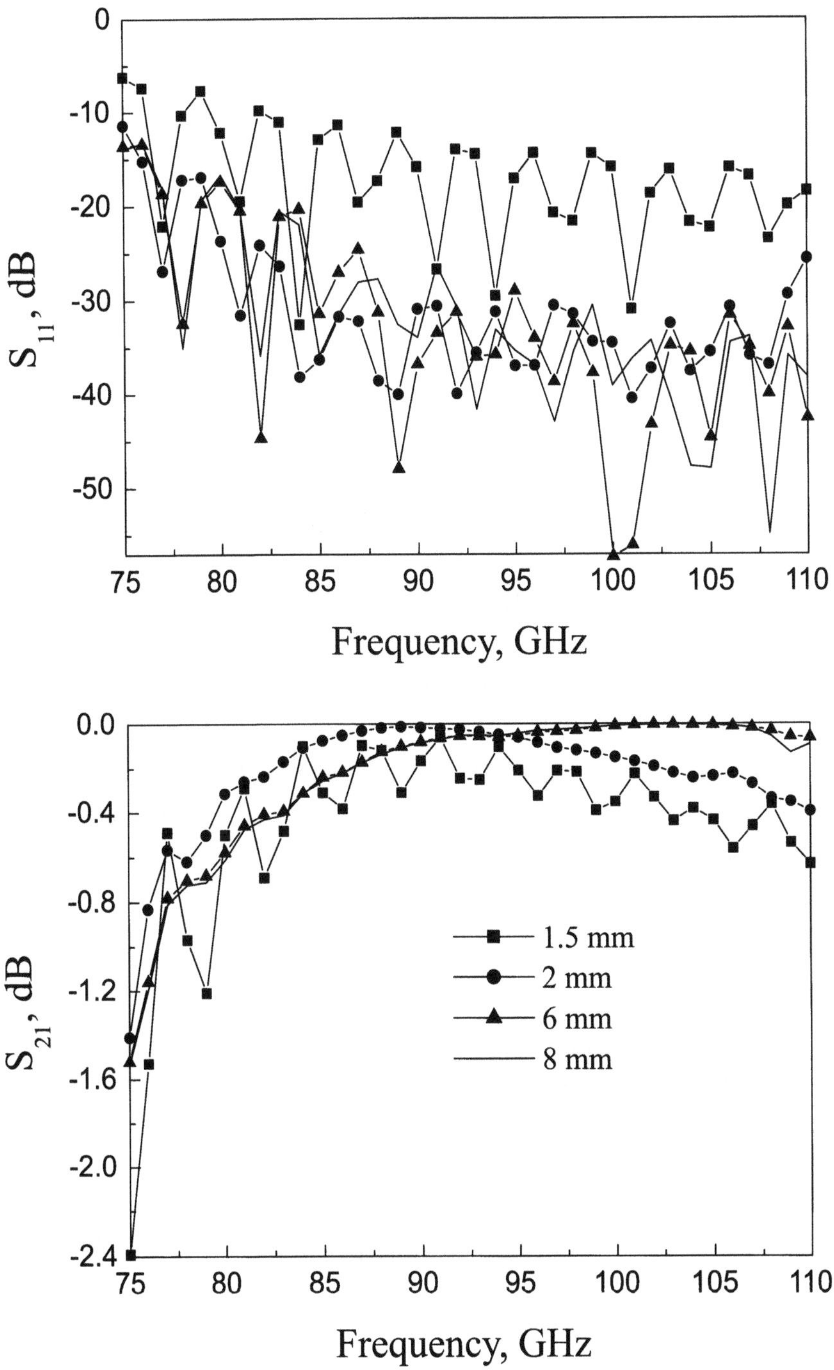

Figure 7.10. Simulation results for symmetrically tapered dielectric rod waveguides of dimensions 1.0×0.5 mm^2 with different taper lengths.

Asymmetrical tapers

Two types of asymmetrical excitation have been simulated: 1) when the axis of the dielectric waveguide is parallel to the axis of the metal waveguide, and 2) when this axis makes an angle of about 4.5° (the angle of the taper is about 9°) (Figure 7.11). It appears that in the second case the transmission coefficient is better than in the first case because in the latter, the mode launcher (the edge of the taper) is too close to the metal wall. The following explains this effect. When the dielectric wedge is symmetrical, the excited field is distributed symmetrically, and the wave transforms to the symmetrical E-mode with approximately the same field distribution. In the case of an asymmetrical taper, a combination of symmetrical and antisymmetrical modes is excited in the wedge, that is, the field distribution does not match the field distribution in the metal waveguide, and the transformation becomes more difficult. The effects of different asymmetrical taper lengths in the range 1 – 8 mm are shown in Figure 7.12.

Influence of anisotropy of the waveguide materials on the transition performance

An interesting effect is found when anisotropy of the waveguide material is introduced in the simulations (Figure 7.13). We have made simulations for a dielectric rod waveguide made of a material with the relative permittivity

$$\varepsilon_r = \begin{pmatrix} 9.39 & 0 & 0 \\ 0 & 9.39 & 0 \\ 0 & 0 & 11.56 \end{pmatrix}$$

corresponding to monocrystal Sapphire with the optical axis directed along the dielectric waveguide. In this case a dip approximately at 105 GHz has been found and also observed experimentally (Section 7.2). Furthermore, one can conclude that an asymmetrical excitation does not significantly corrupt the characteristics. We will further investigate the reasons for these dips in more detail in the case of a symmetrical taper.

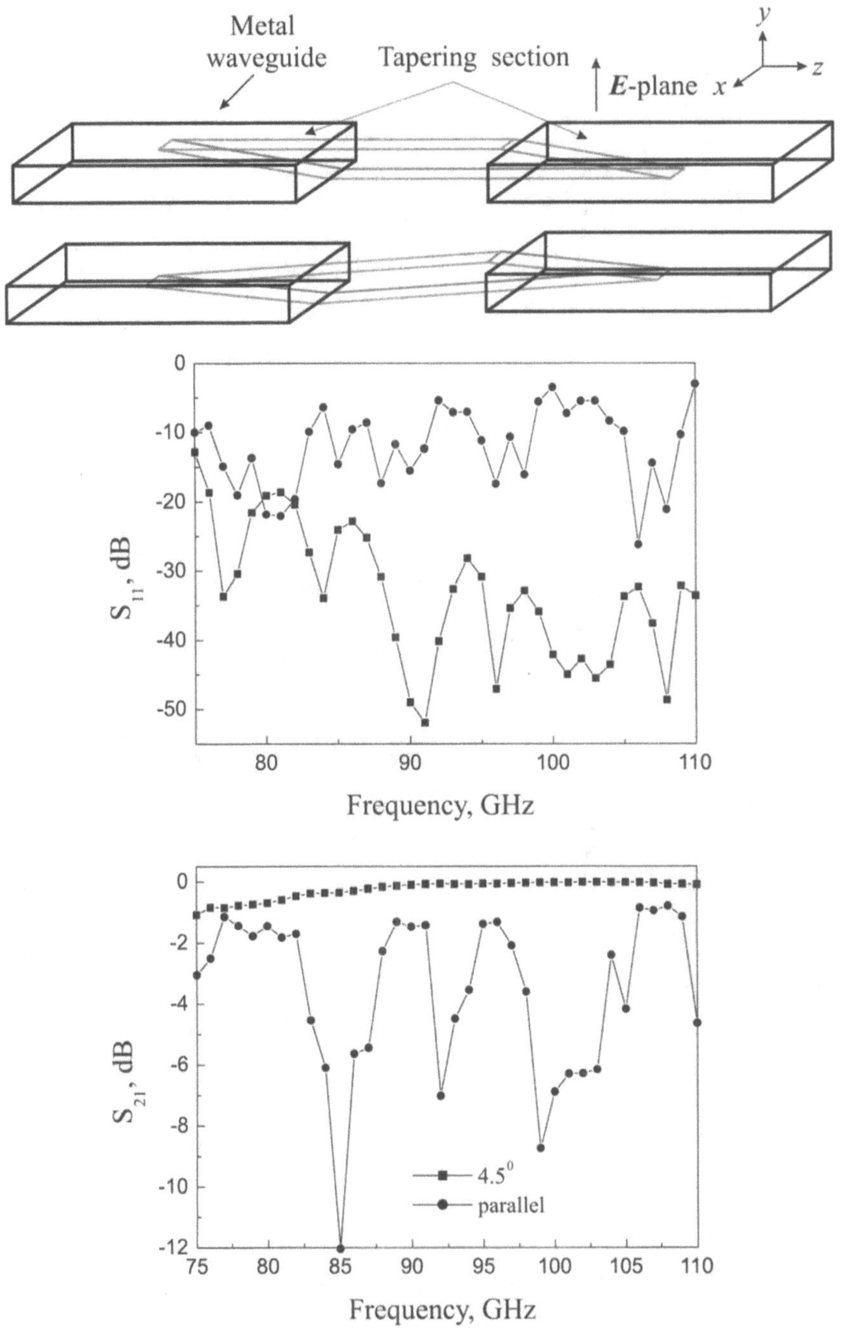

Figure 7.11. Simulation results for an asymmetrical taper of a dielectric waveguide with the dimensions 1.1×0.55 mm^2, and $l = 6$ mm.

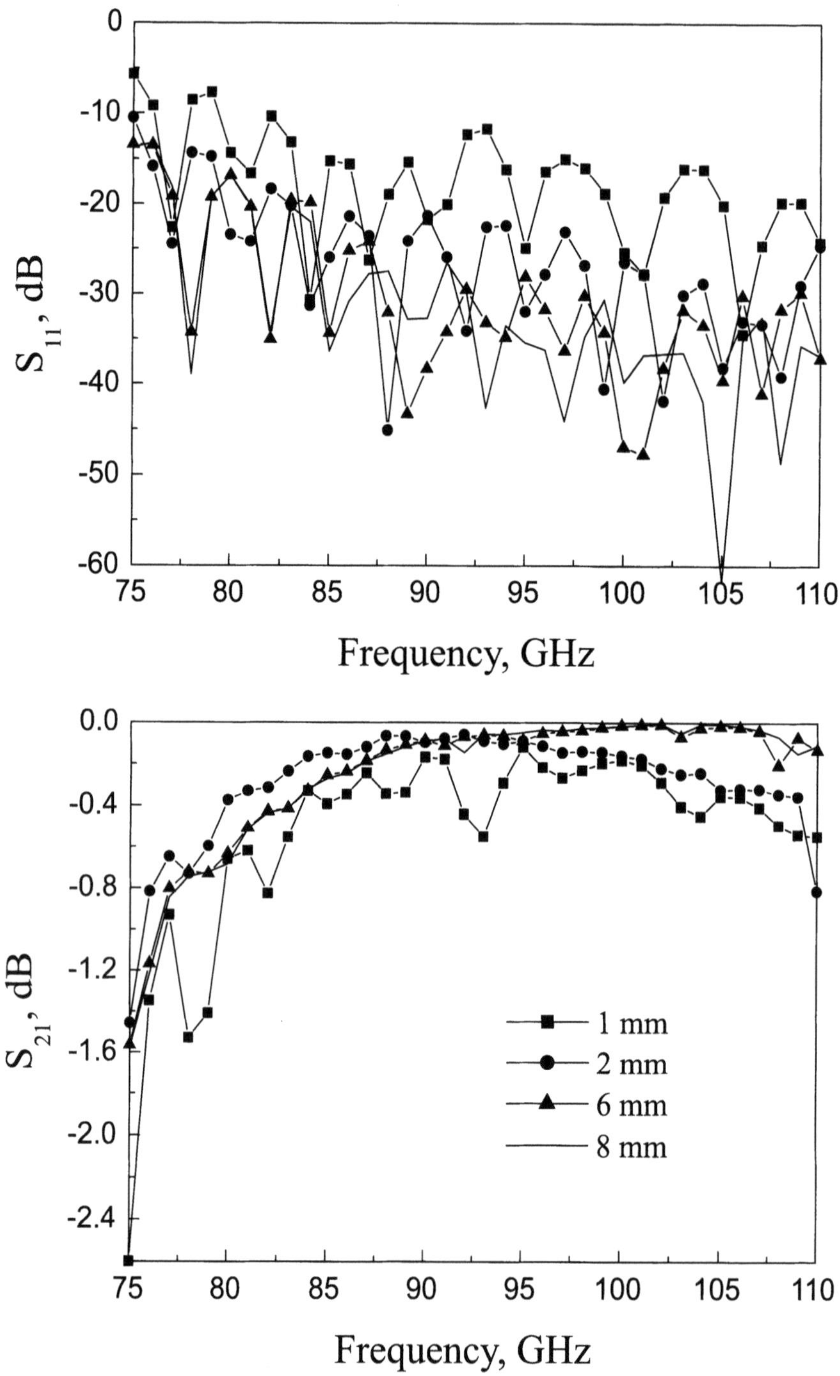

Figure 7.12. Simulation results for dielectric waveguides with 1.0×0.5 mm^2 cross section tapered asymmetrically with different lengths of the tapers.

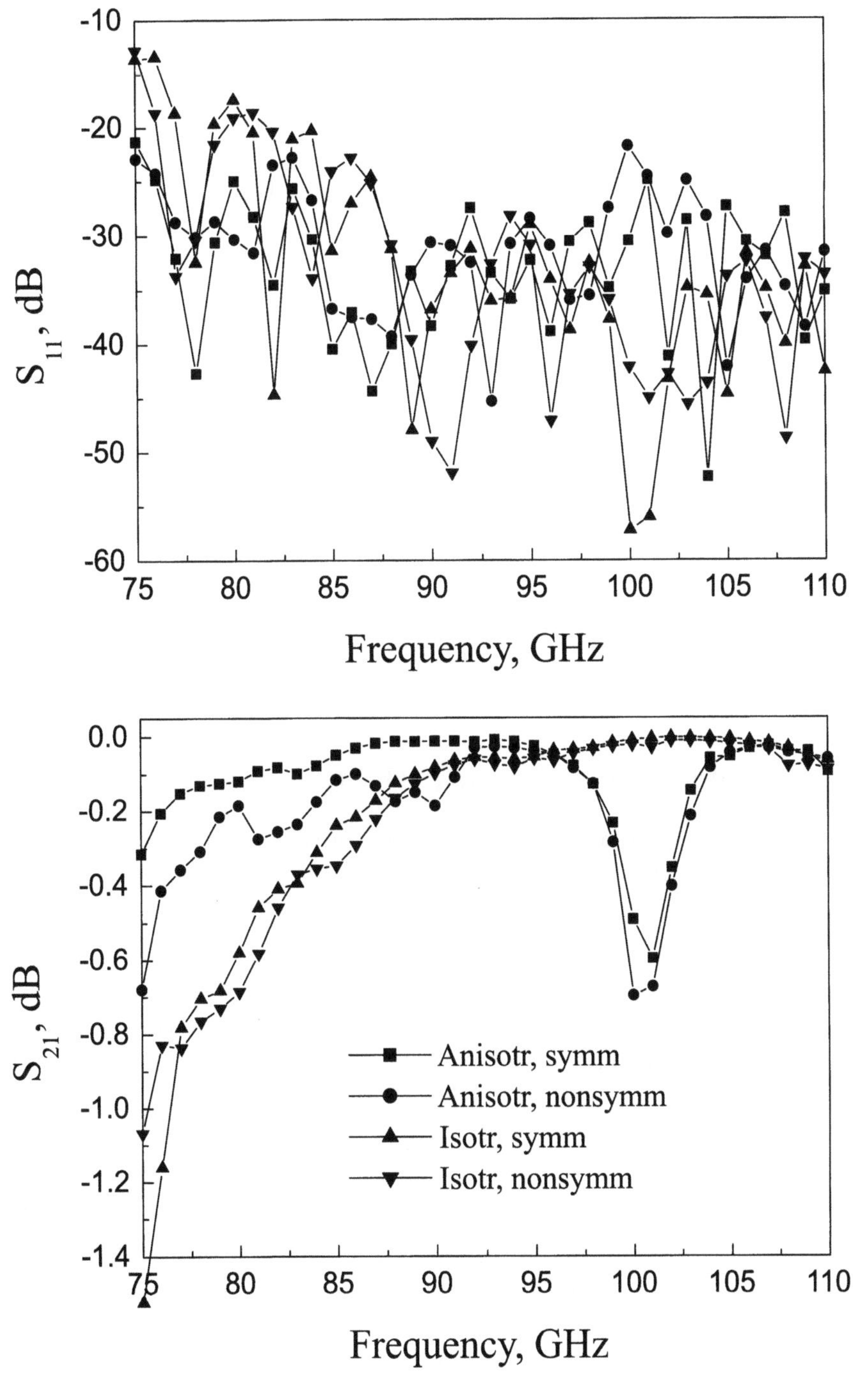

Figure 7.13. Influence of the anisotropy on the transition characteristics ($l = 6$ mm).

Influence of the quality of the taper tip

Two Sapphire waveguide sections with 6 mm tapers (Figure 7.14) have been simulated, and the results are shown in Figure 7.15. As one can see, the 1 mm (about 0.15 mm edge) cleaving increases strongly (extra 10 dB) the reflection but does not change considerably the S_{21} characteristic.

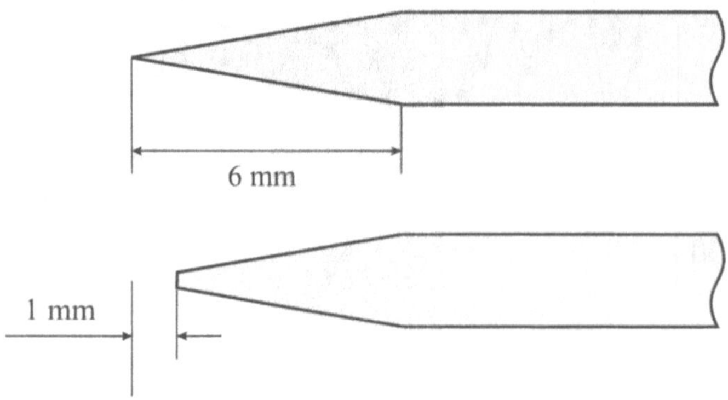

Figure 7.14. Two tapers simulated (with and without cleaving).

7.1.2 Field distribution near the taper section

Let us look in more detail at the field distribution in the dielectric wedge, for example, of the Sapphire dielectric waveguide. In Figure 7.15 dependencies of the insertion losses and reflections on the frequency are shown separately for a symmetrically tapered Sapphire waveguide with the cross section of 1.0×0.5 mm^2 and the taper length of 6 mm. We see, that there is a decrease of the S_{21} parameter with decreasing frequency and a dip with the center approximately at 101 GHz. The E-field distribution at 75 GHz is shown in Figures 7.16 and 7.17. According to the field distribution in the vertical (E) plane we can conclude that in the dielectric wedge two waves are present, "internal" and "external" ones, with different propagation constants (Figure 7.17). We also see that the field near the dielectric surface grows due to concentration of the "external" wave field. On the other hand, this wave couples to the "internal" wave, therefore a maximum in the field strength can be observed. There may be even several maxima, if the coupling between the two waves is small. Also, at 75 GHz the field is weak in a rather large part of the wedge (Figure 7.17). Thus, at lower frequencies the "external" wave is not converting to the "internal" one fast enough.

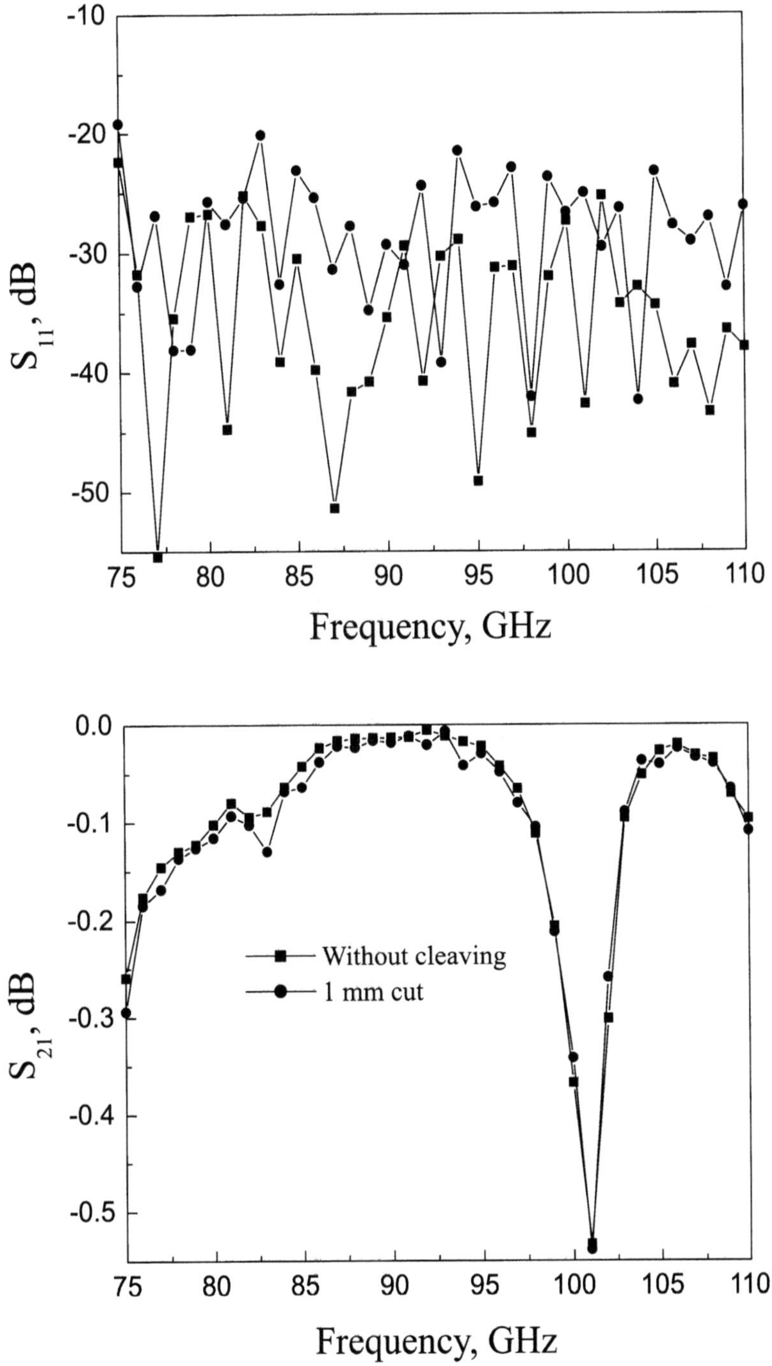

Figure 7.15. Simulation results of a tapered Sapphire waveguide with and without 1 mm cleaving.

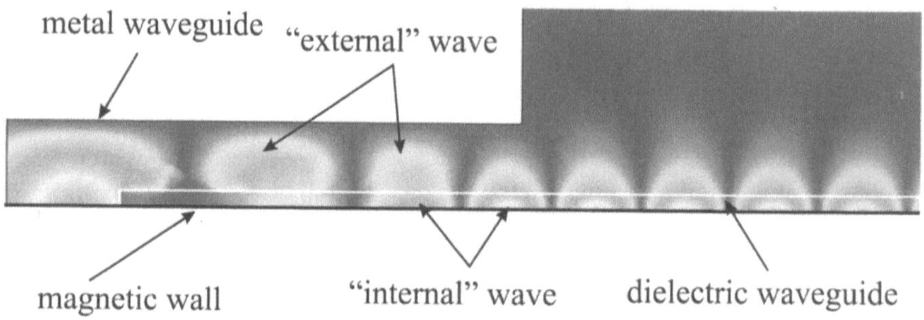

Figure 7.16. Electric field strength in a Sapphire wedge at a particular time instance. The signal frequency is 75 GHz. The horizontal (H) plane.

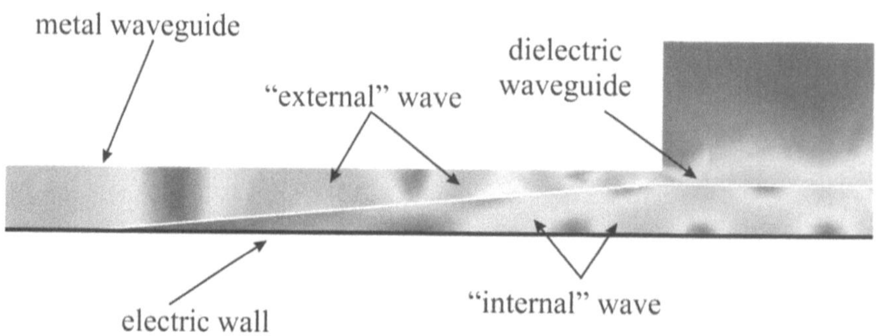

Figure 7.17. Electric field distribution in a Sapphire wedge at a particular time instance. The signal frequency is 75 GHz. The vertical (E) plane.

In Figure 7.18 one can see that several field maxima are present on the surface of the wedge. One of them is at the end of the metal waveguide. This results in additional radiation into the environment. As we see in Figure 7.15, the best characteristics for this Sapphire waveguide section are obtained approximately at 90 GHz, therefore let us look at the field distribution at this frequency. In Figure 7.19 we can see, that the field is concentrated stronger due to a higher frequency. Also, there is a field maximum at the surface of the wedge. In Figure 7.20 the electric field distribution at some instance of time is shown, where one can see again "internal" and "external" waves.

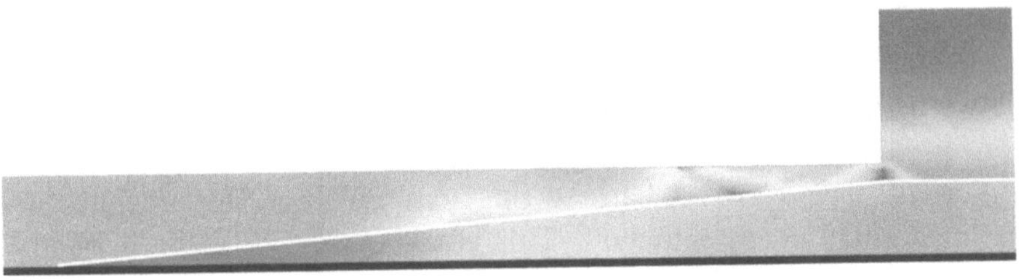

Figure 7.18. Amplitude distribution of the electric field in a transition from a metal waveguide to a Sapphire dielectric rod waveguide. The vertical (E) plane. The frequency is 75 GHz.

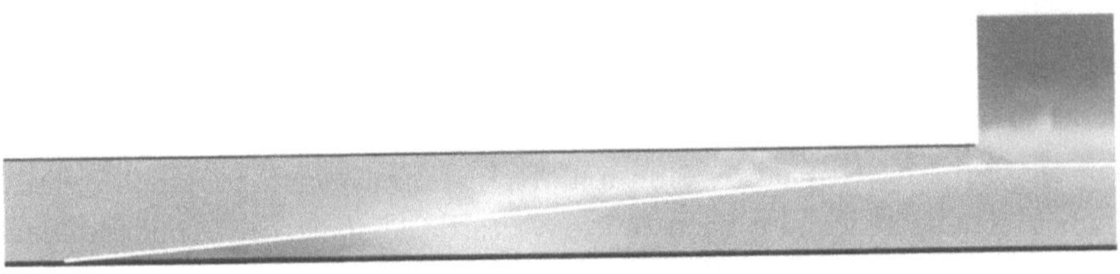

Figure 7.19. Amplitude distribution of the electric field in a transition from a metal waveguide to a Sapphire dielectric rod waveguide. The vertical (E) plane, and the frequency is 90 GHz.

The case of the frequency of 101 GHz is more interesting. At 101 GHz

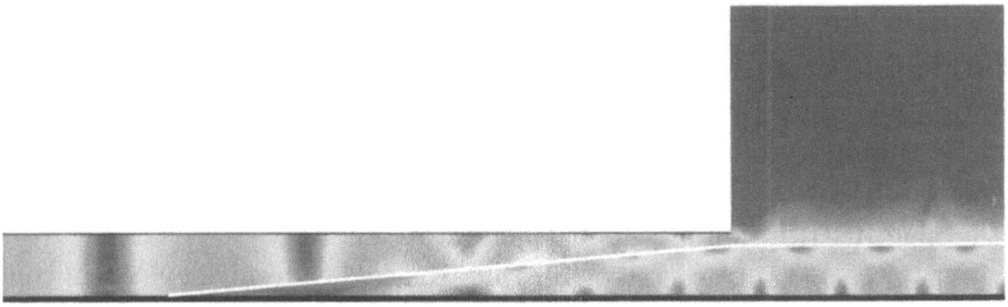

Figure 7.20. Electric field distribution in a transition from a metal waveguide to a Sapphire dielectric rod waveguide at a particular time instance. The vertical (E) plane. The signal frequency is 90 GHz.

(Figure 7.21) one can observe the following phenomena. The "external" wave gradually converts to the "internal" one, but the phase difference between these waves increases. As a result, at a particular place the converted wave is out of phase with the "internal" wave, that is, the "internal" wave converts back to the "external" one (Figure 7.21) and radiates into the environment. This is the reason for a dip in the S_{21} characteristic at the frequency of 101 GHz.

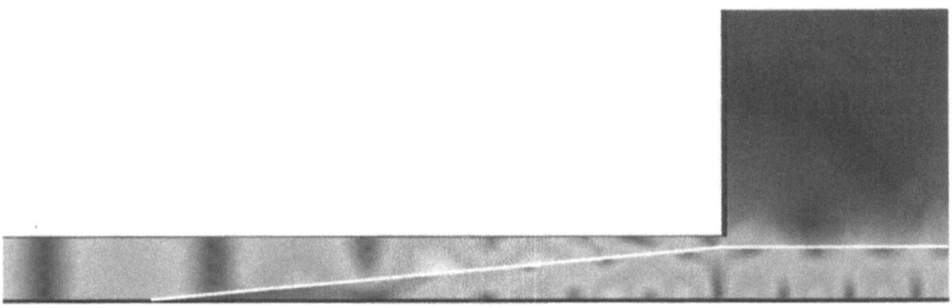

Figure 7.21. Electric field distribution in a metal waveguide to Sapphire DRW transition at a particular time instance. Vertical (E) plane, the frequency equals 101 GHz.

Thus, at 101 GHz the main mechanism of losses is the conversion of the "internal" wave to the "external" one followed by radiation to the environment. To prevent this, the phase difference between the "internal" and "external" waves should not change very much along the taper length.

7.2 EXPERIMENTAL MEASUREMENTS OF DIELECTRIC WAVEGUIDES

High resistivity (> 10 kΩ/cm^2) Silicon, semi-insulated GaAs with carrier concentration of $n \sim 10^8$ cm^{-3}, monocrystal Sapphire, Boron Nitride, ferrites and other materials with tan$\delta \leq 10^{-4}$ can be used as materials for fabrication of dielectric rod waveguides.

7.2.1 Waveguide samples and the experimental setup

Monocrystal Sapphire cut along the optical axis looks very attractive for dielectric waveguide fabrication. Cross section dimensions of 1.0×0.5 mm^2 with a 6 mm taper length have been chosen for the frequency range of $75 - 110$ GHz according to the results of HFSS simulations. An asymmetrical taper section has been chosen, because of easier fabrication. The length of the tapered section is 6 mm, so that the angle of the taper is about 9°. Four samples with 47 mm length and four samples with 110 mm length have been measured, all with the 1.0×0.5 mm^2 cross section.

GaAs dielectric rod waveguides oriented along the [110] direction have been cut from a semi-insulating (100) GaAs wafer with the electron concentration $n \sim 10^8$ cm^{-3}, $\tan\delta \sim 3 \cdot 10^{-4}$, and $\varepsilon \sim 13.3$. The dielectric waveguides have been only cut without polishing, therefore the taper sections are not perfect and may have small shape defects. The GaAs dielectric waveguide samples are described in Table 7.1.

Table 7.1. GaAs dielectric waveguide samples

Sample number	Cross section	Total length	Taper
GaAs DRW 1	1.04×0.51 mm^2	45 mm	6 mm sym
GaAs DRW 2	1.01×0.51 mm^2	45 mm	6 mm asym
GaAs DRW 3	1.43×0.51 mm^2	49 mm	8 mm sym
GaAs DRW 4	1.43×0.51 mm^2	48 mm	8 mm asym

The taper plane was chosen to be in the E-plane of the metal waveguide according to the results of Section 7.1.1. The dielectric waveguide was supported by a styrofoam holder. The transition from the standard metal waveguide to the dielectric one and vice versa is shown in Figure 7.22. A vector network analyzer HP 8510 was used to measure the S-parameter characteristics.

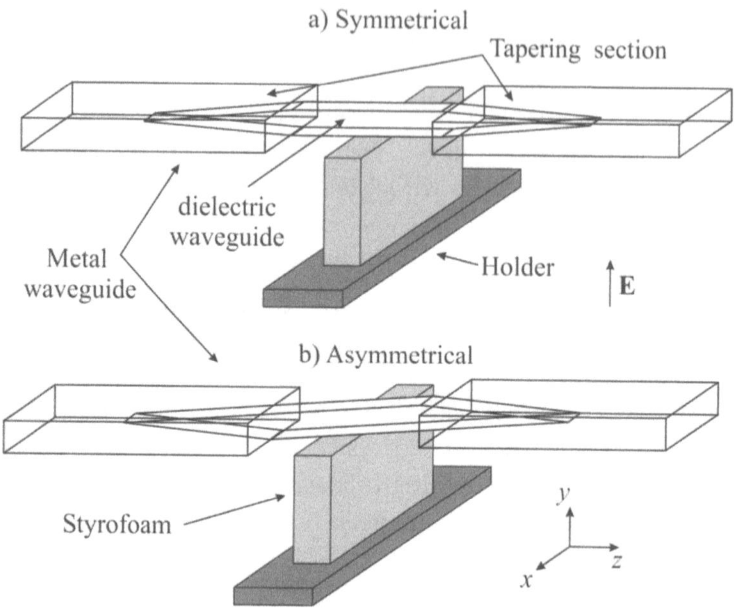

Figure 7.22. Experimental setup for measuring the characteristics of transitions from a metal waveguide to dielectric waveguides: a) symmetrical case, b) asymmetrical case. The metal waveguides are connected to a vector network analyzer.

7.2.2 Sapphire dielectric rod waveguides

The insertion losses for 47 mm and 110 mm Sapphire waveguides (asymmetrical taper in both cases, see Figure 7.22b) are presented in Figure 7.23. We see that the 47 mm Sapphire waveguide section has the insertion loss of 0.1 – 0.3 dB at frequencies 77 – 98 GHz. One can see that from the frequency 98 GHz upwards the losses increase and reach a maximum at 105 GHz. Such undesirable corruption of the characteristics can be explained by difficulties of the excitation occurring due to non-idealities of the taper sections, and also due to a possible excitation of higher-order modes and "internal" and "external" waves (see Section 7.1.2).

A typical S_{11} curve for 47 mm and 110 mm long Sapphire dielectric waveguide sections with two transitions to metal waveguides is shown in Figure 7.24. The maximum value of VSWR for two transitions is 1.23 ($\approx$1.13 for one transition), which corresponds to 0.07 dB loss. At the lower frequency end the insertion losses are higher, and analyzing

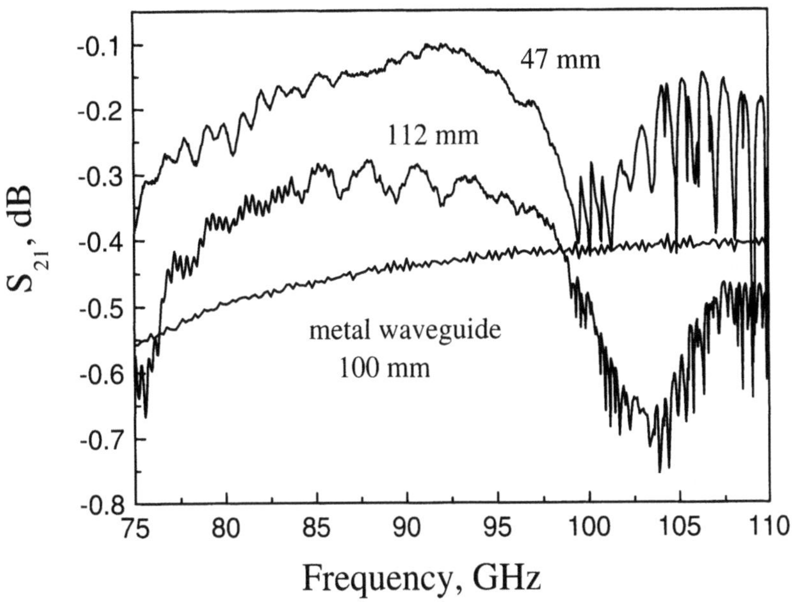

Figure 7.23. S_{21} (absolute values) of 47 mm and 112 mm sections of a Sapphire dielectric waveguide.

Figure 7.23 we conclude that the radiation loss is the dominant factor at the lower frequency end. In the middle band, however, the absorption loss is the main factor. The ripples of the S_{11} characteristics are caused by the interference between the ends of the dielectric waveguide section.

Removing the receiving metal waveguide results in an increase of the reflection coefficient approximately by 10 dB (Figure 7.25). This means that while such a structure results in a well-matched dielectric rod antenna, the antenna is not that well matched to free space as the dielectric waveguide is matched to another metal waveguide.

A Sapphire dielectric waveguide with 47 mm length and 1.0×0.5 mm^2 cross section can be also used as an antenna. The corresponding experimental setup is shown in Figure 7.26. A Sapphire waveguide is inserted into a Styrofoam holder that is attached to a standard metal waveguide using flange pins. Similarly, a GaAs waveguide antenna was constructed.

Radiation pattern measurements were carried out with an antenna rotator and AB Millimetre 8-350 vector network analyzer in both E- and

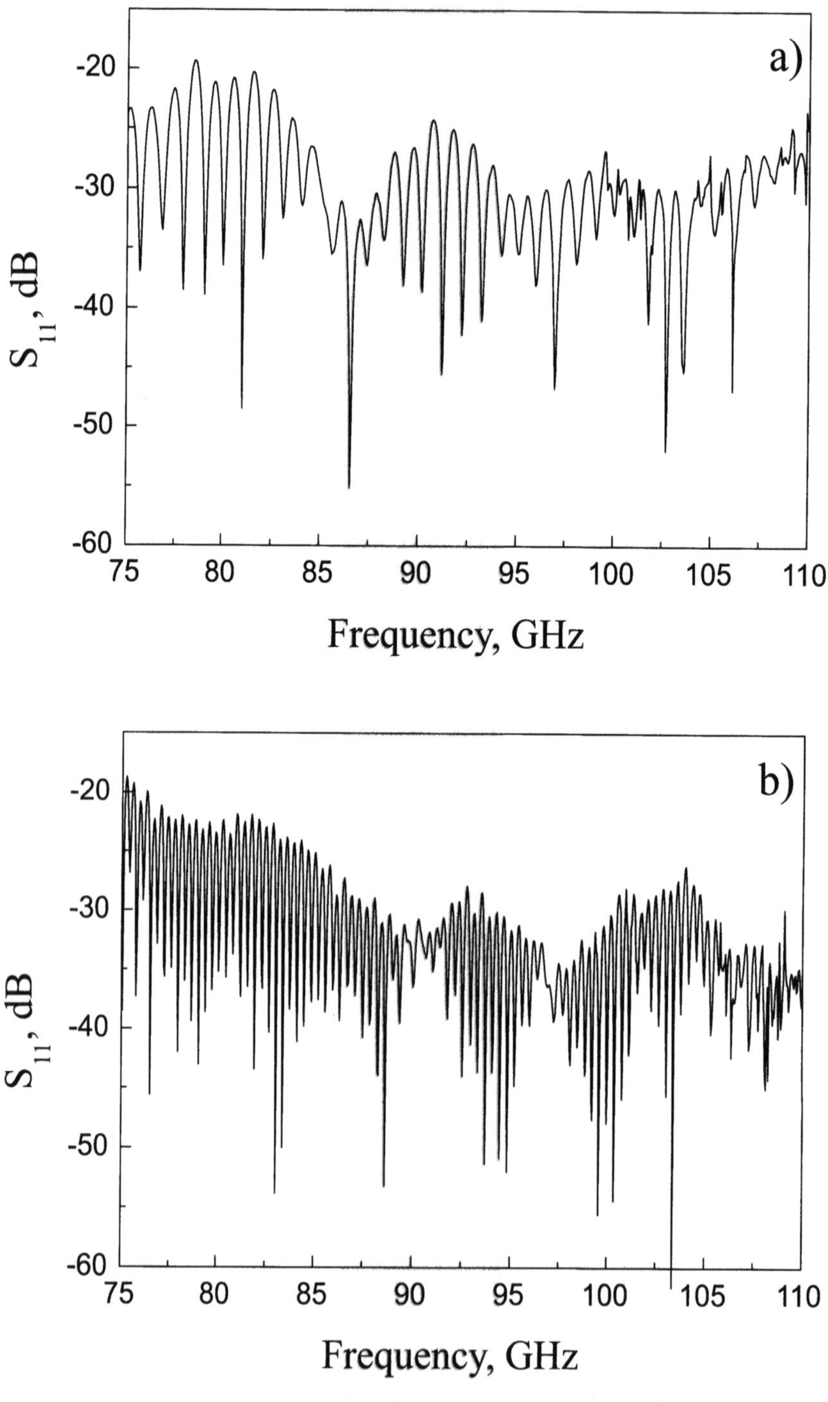

Figure 7.24. S_{11} for 47 mm (a) and 112 mm (b) long Sapphire dielectric rod waveguide sections.

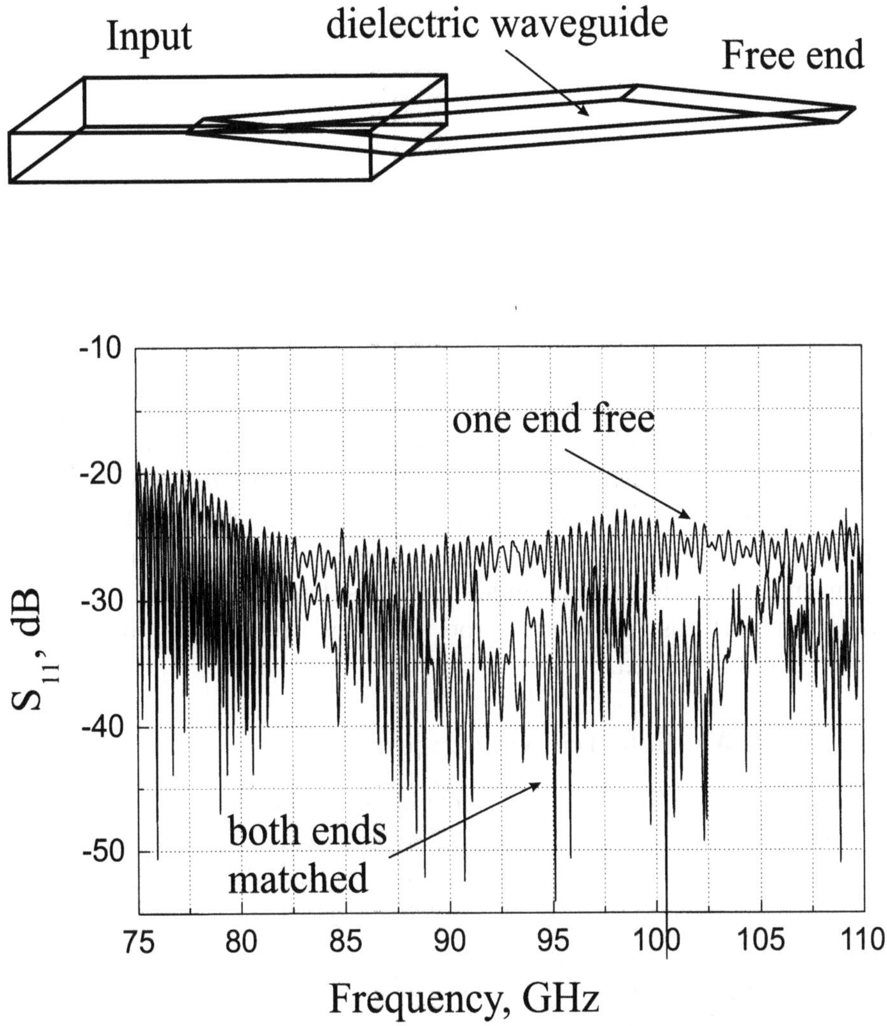

Figure 7.25. Measured results for a dielectric waveguide with an open and matched end.

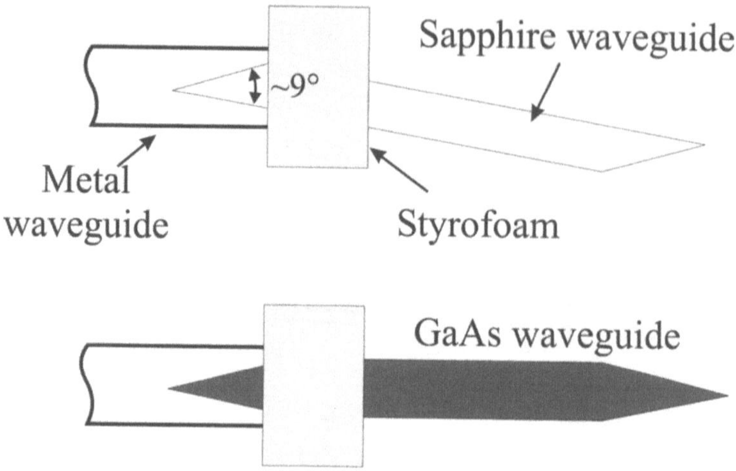

Figure 7.26. Schematic structure of the dielectric rod antennas.

H-planes and the results are shown in Figures 7.27 and 7.28, respectively. The −3 dB beamwidths of the antennas are about 65–70 degrees in both E- and H-planes. The radiation patterns are somewhat similar to that of an open end of a metal waveguide. However, the reflections from an open end of the metal waveguide are considerably higher. Moreover, radiation patterns of dielectric rod antennas are rather stable in the whole frequency band.

7.2.3 GaAs dielectric waveguides

Four different GaAs dielectric waveguide samples of the length 45 – 49 mm have been measured (Figure 7.29). Typically, the absolute value of S_{21} is approximately $0.4-0.5$ dB, while S_{11} is mainly below 30 dB. One can see that asymmetrically tapered dielectric waveguides have worse characteristics at both ends of this frequency range. Dielectric waveguides with larger vertical dimensions ($\approx 1.5 \times 0.5$ mm^2 and 8 mm taper) also have good characteristics. The reflections are lower than for Sapphire dielectric waveguides. The characteristics at frequencies above 100 GHz are corrupted similarly to those of Sapphire waveguides.

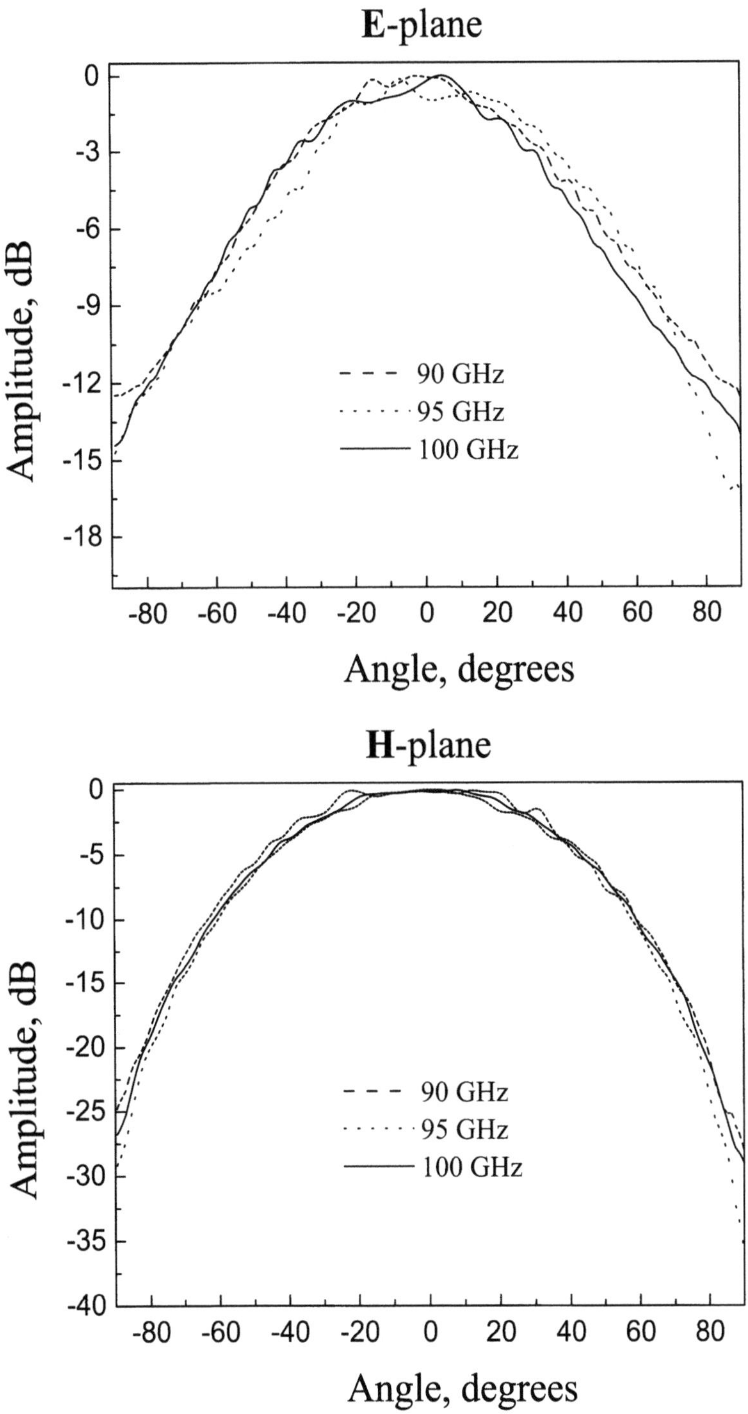

Figure 7.27. Radiation patterns of Sapphire dielectric waveguide antenna.

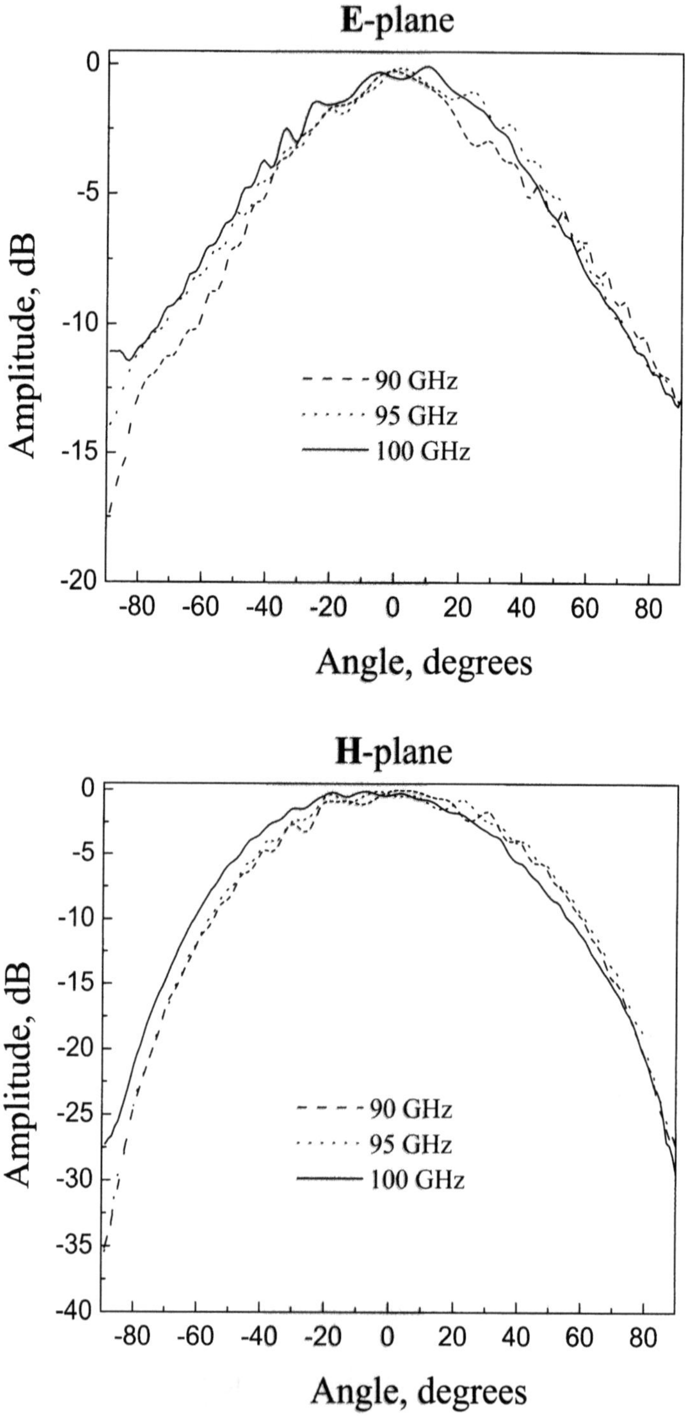

Figure 7.28. Radiation patterns of GaAs dielectric waveguide antenna.

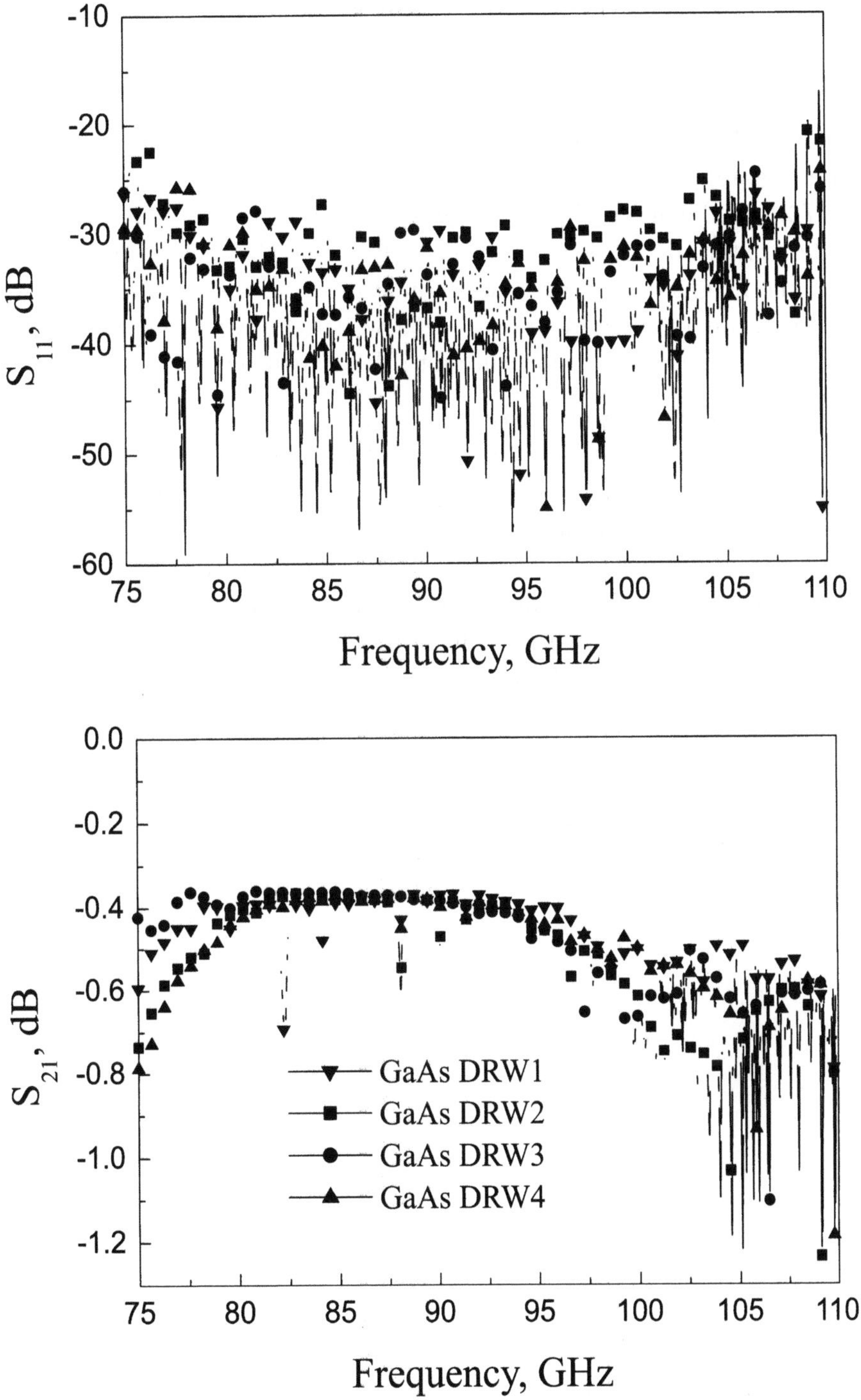

Figure 7.29. Measured results for a dielectric waveguide with an open end.

7.2.4 Horn-like structure implementation

An undesirable dip in the S-parameter characteristic at frequencies around 100 GHz has been observed both with HFSS simulations and experimentally. Analyzing the filed distributions images (e.g., Figure 7.21) one can see that it is related to a nonuniform transformation of the "external" wave to the "internal" one. The field spreads wider than the height of the metal waveguide in the E-plane resulting in extra radiation into environment. One way to prevent this effect is to introduce a "horn"-like structure into the transition.

HFSS simulation were carried out in order to investigate the influence of a "horn"-like structure. The "horn"-like structure (Figure 7.30) has been simulated with different flare angles and lengths in the E-plane. In Figure 7.31 the simulation results for flare angles of 3°, 10°, 40°, 60°, and lengths d of 0.3 mm, 0.5 mm, and 0.7 mm are shown. One can see that in spite of small dimensions of the "horn" both the flare angle and its length affect the transitions characteristics and allow to suppress the dip in the transmission characteristic.

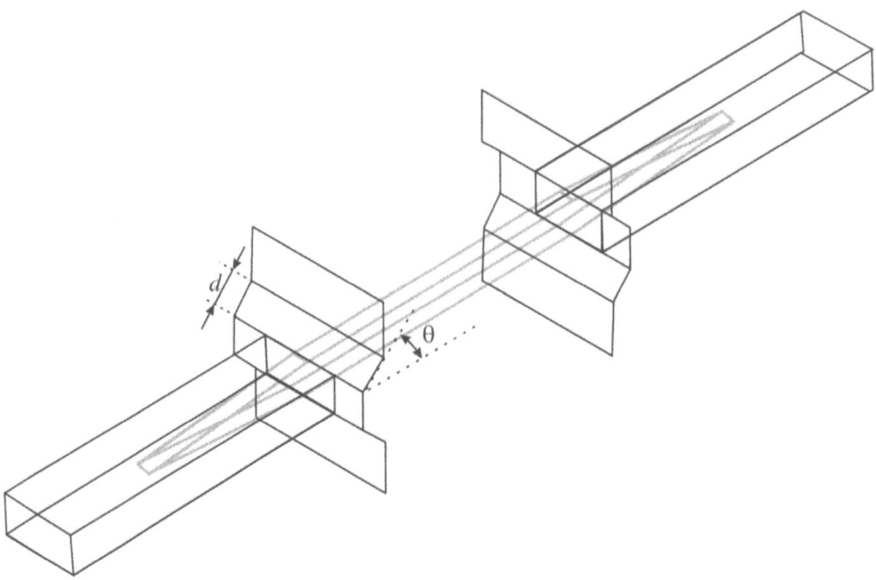

Figure 7.30. A "Horn"-like structure introduced into simulations.

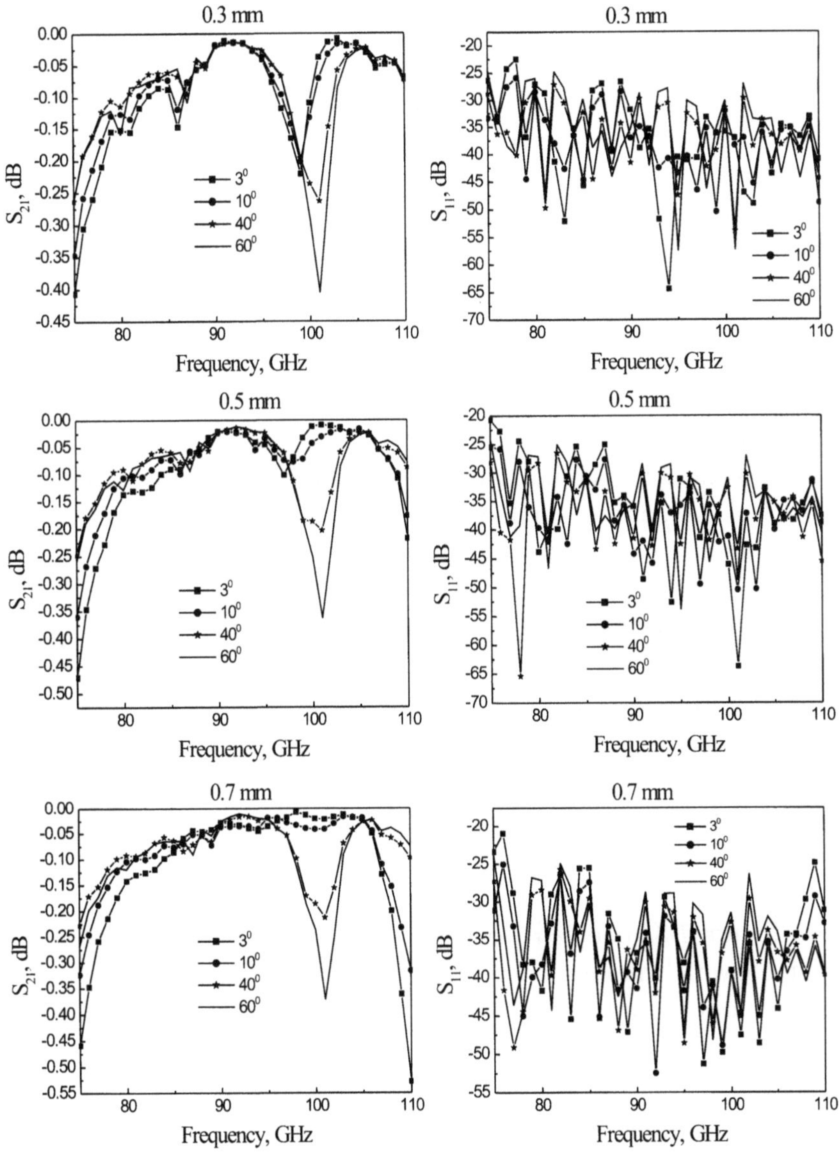

Figure 7.31. HFSS simulation results for a "horn"-like structure.

7.2.5 Conclusions

The experimental results indicate that rectangular monocrystalline Sapphire and GaAs dielectric rod waveguides can be well matched with a single-mode metal waveguide, and they have low insertion loss in the 75 – 110 GHz frequency range. The insertion loss of Sapphire waveguides is very low at frequencies 77 – 98 GHz.

Excitation of Sapphire dielectric waveguides with an asymmetrical taper section (asymmetric relative to the optical axis) and excitation of GaAs dielectric waveguides with both symmetrical and asymmetrical taper sections have been experimentally investigated. The tapers have been made in the E-plane of the metal waveguide. The insertion losses are mainly caused by the transition from metal waveguides to Sapphire waveguides and vice versa (radiation) at the lower frequency end and by absorption in the dielectric waveguide itself in the middle of the frequency range. Thus, we can conclude that the materials with a relatively high dielectric constant and a low loss tangent, such as monocrystal Sapphire, Si, GaAs, and so forth, can be successfully employed as dielectric waveguides at frequencies above 75 GHz.

7.3 SOME NOTES ABOUT METAL WAVEGUIDES

At millimeter-wave frequencies measurements usually have to be carried out more carefully than at microwave frequencies because of higher requirements on tolerances. For example, flange connectors are manufactured with two pins, and metal waveguides are plated with gold or silver. These measures are not so necessary at the microwave frequencies.

In Figure 7.32 one can see measured results obtained with HP8510 vector network analyzer. Here, the dotted line corresponds to a direct connection of the input and output waveguides. The dashed lines have been obtained when measuring the characteristics of a 100 mm long gold-plated WR-10 (2.54×1.27 mm^2) metal waveguide section. To estimate the flange losses, two shorter waveguide sections were connected with approximately the same total length. Looking at the dash-dotted line, one can see that an additional flange connection results in approximately 0.15 dB extra loss. Also, it does not improve the reflection characteristics. However, the flange losses strongly depend on the quality of flanges and might be different in case of other waveguide sections.

Comparing the losses for a longer and a shorter sapphire waveguide

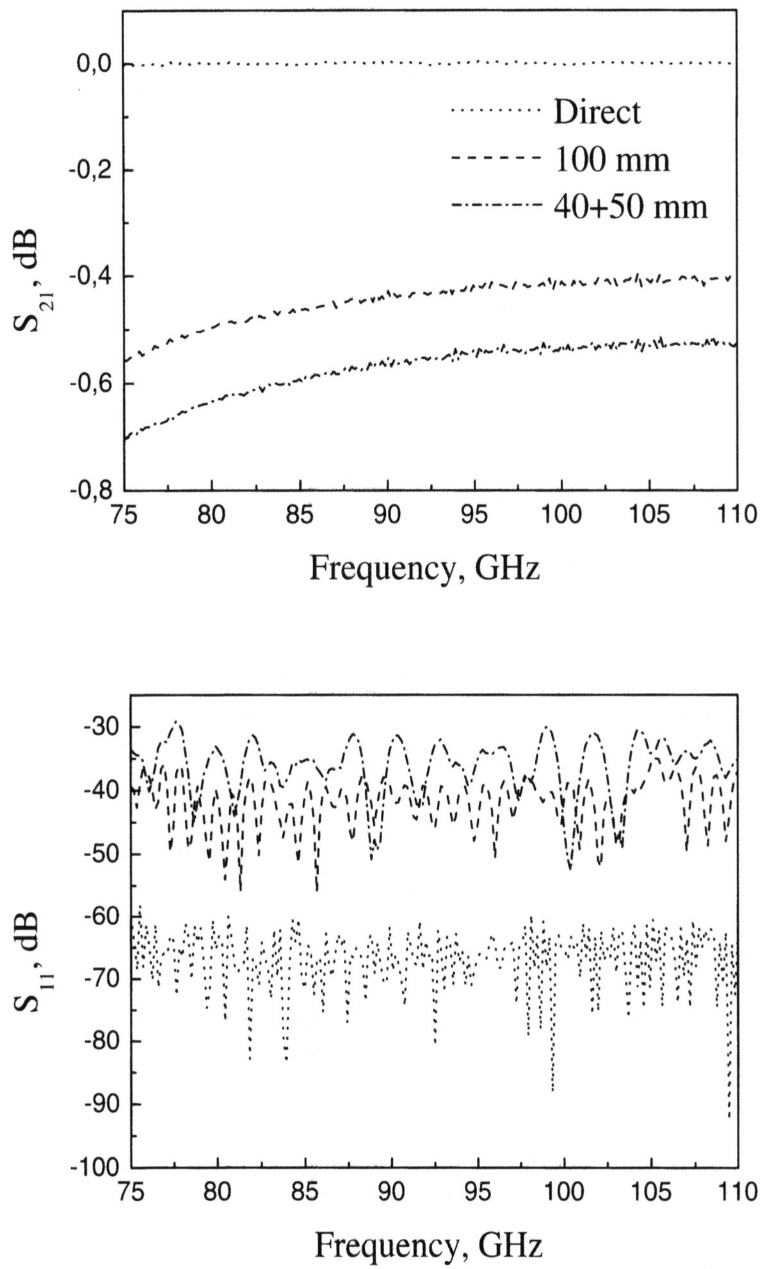

Figure 7.32. Insertion losses and reflections in metal waveguide sections consisting of one and two pieces.

sections in Figure 7.23, one can conclude that the transition losses from a metal to a dielectric waveguide might be somewhat lower than the losses at a typical flange connection between two metal waveguides.

In high-power applications, flange connections can be overheated. Dielectric insertions are used to decrease this effect.

References

[1] T.N. Trinch and R. Mittra, Transitions from metal to dielectric waveguide, *Microwave J.*, Nov. 1980, pp. 71-74.

[2] J.A. Paul and Y.W. Chang, Millimeter-wave image-guide integrated passive devices, *IEEE Transactions on Microwave Theory and Techniques*, vol. MTT-26, no. 10, 1978, pp. 751-754.

[3] H. Jacobs and M. Chrepta, Electronic phase shifter for millimeter-wave semiconductor dielectric integrated circuits, *IEEE Transactions on Microwave Theory and Techniques*, vol. MTT-22, no. 4, 1974, pp. 411-417.

[4] V.V. Meriakri and M.P. Parkhomenko, Millimeter-wave dielectric strip waveguides made of ferrites and phase shifters based on these waveguides, *Electromagnetic Waves & Electronic Systems*, vol. 1, no. 1, 1996, pp. 89-96.

[5] V.V. Meriakri, B.A. Murmuzhev, and M.P. Parkhomenko, Millimeter wave devices based on dielectric, ferrite and semiconductor waveguides, *Proc. of the Microwave and Optoelectronics Conference*, vol. 2, 1997, pp. 431-433.

[6] J. Weinzierl, Ch. Fluhrer, and H. Brand, Dielectric waveguides at submilliter wavelengths, *Proc. of the IEEE Sixth International Conference on Terahertz Electronics*, 1998, pp. 166-169.

[7] D.V. Lioubchenko, S.N. Dudorov, and A.V. Räisänen, Development of rectangular open dielectric waveguide sections for the frequency range of 75-110 GHz, *Proc. of the 31st European Microwave Conference*, vol. 2, 2001, pp. 201-204.

Chapter 8

Dielectric waveguide devices and integrated circuits

Dielectric waveguides for millimeter-wave applications offer advantages of lower cost, easier manufacturing, higher tolerances, etc., and they are actively studied nowadays. After the regular dielectric rod waveguide itself had been understood, the development of integrated circuits based on such waveguides started. At millimeter waves, rectangular dielectric waveguiding structures are more attractive because of their better compatibility with integrated circuits and easier manufacturing processes. Single-mode dielectric waveguiding structures and devices based on them are in principle similar to the conventional microwave microstrip lines and circuits. However, when the frequency increases, the use of microstrip transmission lines becomes difficult. For example, at 50 GHz a microstrip line on an Alumina substrate has losses of 57 dB/m [1, 2], while for dielectric waveguides several dB/m is typical (in [3], 3.76 dB/m has been reported). The existing dielectric waveguide circuits can be classified as belonging to two main types: devices based on a non-radiative waveguide (dielectric waveguide between two metal plates), and devices based on a planar waveguide (dielectric waveguiding structures on dielectric substrates).

8.1 DIELECTRIC WAVEGUIDES FOR INTEGRATED CIRCUITS

We begin with a discussion of various dielectric waveguides that find applications in millimeter-wave integrated circuits.

8.1.1 Non-radiative dielectric waveguide

The non-radiative dielectric (NRD) waveguide was first proposed in [3]. It is known that if two parallel metal plates are separated by a distance smaller than one half of the wavelength, the electromagnetic wave, polarized so that the electric field is parallel to the plates, can not propagate. However, if a dielectric strip is inserted between the metal plates, such a wave can propagate (Figure 8.1). To prevent excitation of the ortho-

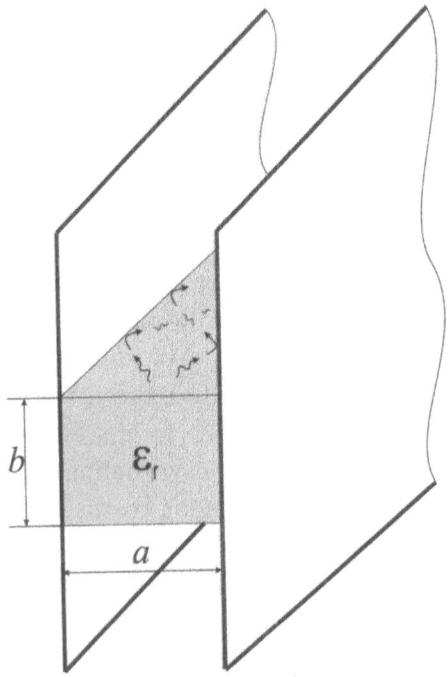

Figure 8.1. Geometrical structure of a non-radiative dielectric waveguide.

gonal polarization, a simple mode suppressor can be introduced, such as described in [2].

The non-radiating dielectric waveguide can be imagined as a single-mode rectangular metal waveguide filled with a dielectric, with two wider walls removed and enlarged narrower walls. Thus, the ohmic losses in metal are significantly reduced. Also, radiation can be almost completely suppressed, and the design of such devices as directional couplers, power dividers, circulators, etc., can be considerably simplified. The operational mode used in the non-radiating dielectric waveguide can be imagined as a combination of two TM mode waves of a dielectric slab waveguide, reflecting from the metal plates.

As an example, the non-radiative nature has been illustrated in [2] (Section 3), when the reflection of the truncated end of a non-radiative dielectric waveguide was observed. The electromagnetic wave was reflected almost completely. For the design of the NRD waveguide the following recommendations are given in [2]:

$$\frac{a}{\lambda} \approx 0.45 \tag{8.1}$$

$$\sqrt{\varepsilon_r - 1}\,\frac{b}{\lambda} \approx 0.4\ldots 0.6 \tag{8.2}$$

As was shown in [3], the frequency region of the single-mode operation of the non-radiative dielectric waveguide becomes larger when the permittivity of dielectric strip increases, which also allows one to reduce the device size.

Devices based on the non-radiating dielectric waveguide can have very good performance. For example, in [3] it was demonstrated that a 90° bend can have the insertion loss as low as 0.3 dB, while for a 180° bend after adjustment the loss even could not be measured.

Different filters consisting of one or more sections of non-radiating dielectric waveguides and several dielectric disks were proposed in [4]. Other devices are reviewed in [2], such as matched terminator, directional coupler, circulator, beam lead diode mount, Gunn diode oscillator, power divider, leaky-wave antenna. Also, a transmitting and receiving modules using these components are reviewed there.

However, the non-radiative dielectric waveguide has some difficulties in the use in integrated circuits, especially for integrating three-terminal devices, such as transistors. Also, there is a fundamental limitation on the distance between plates. An effort to connect planar circuits (based on microstrip lines) and the non-radiative dielectric waveguide technology together has been made in [5].

8.1.2 Dielectric waveguide circuits on metal and dielectric substrates

This approach has been widely employed for a long time in the infrared and the optical frequency ranges (so-called "integrated optics" [6]). Although this technology was developed earlier than the non-radiative dielectric guides, there are fewer publications on millimeter-wave integrated circuits based on dielectric waveguides on metal or dielectric substrates.

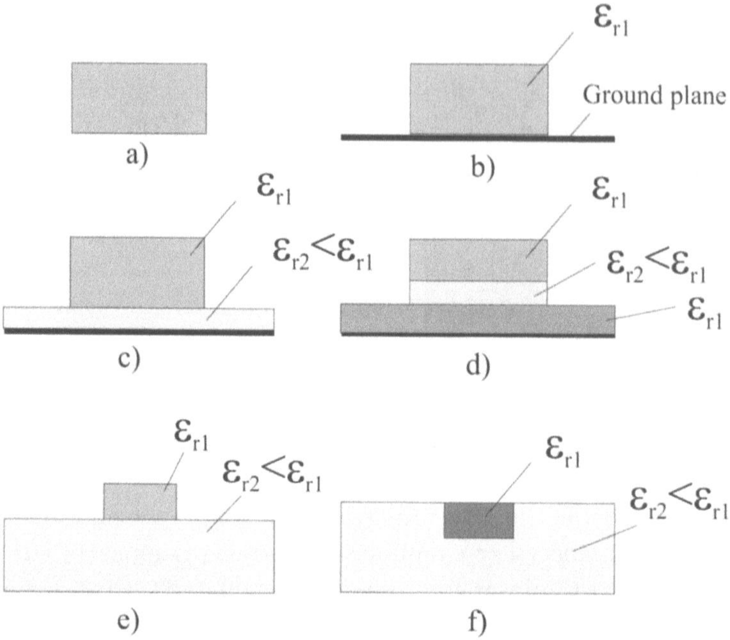

Figure 8.2. Main types of dielectric waveguiding structures: a) dielectric rod waveguide, b) image waveguide, c) insulated image waveguide, d) layered ribbon waveguide, e), f) other types.

The main existing types of dielectric waveguiding structures are shown in Figure 8.2. The ground metal plane in the image and insulated image waveguides makes them applicable in active circuits. The insulated image waveguide differs from the image waveguide by an additional layer with the permittivity less than that of the strip. As the power propagates mainly in the rod, the field near metal is reduced and thus the metal losses decrease. The waveguides in Figure 8.2, e) and f) have even lower losses due to the absence of metal layers. A review of different millimeter waveguiding structures and description of some devices can be found e.g., in [7]. Also a description of a V-band receiving module can be found there with a nice picture where one can see a dielectric waveguide circuit. Another review can be found in [8].

A simple method for calculation of rectangular dielectric waveguide structures is proposed in [8] (the method of effective dielectric constant).

Also, two other interesting structures have been proposed: the strip dielectric waveguide and the inverted strip dielectric waveguide (Figure 8.3). The difference of these structures is that the permittivity of

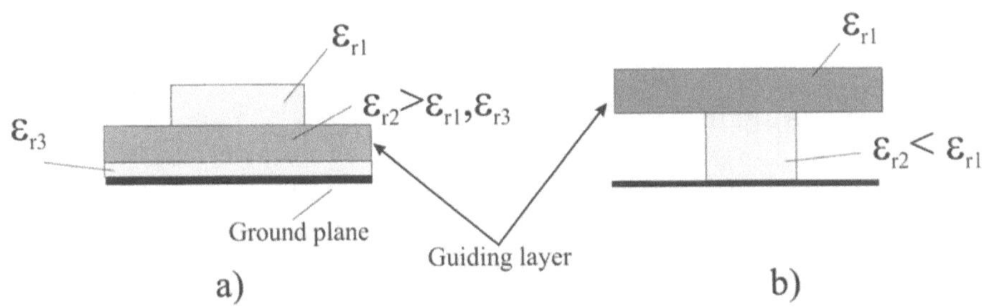

Figure 8.3. Other types of dielectric waveguiding structures: a) strip dielectric waveguide, b) inverted strip dielectric waveguide [10].

the dielectric strip is chosen to be smaller than that of the so-called guiding layer. The electromagnetic wave propagates mainly in the guiding layer which has the largest permittivity, while the dielectric strips with a smaller dielectric constant are used to produce a "lens effect" [10]. Results of the field distribution measurements are also presented in the same paper.

The main advantage of the inverted strip waveguide (Figure 8.3b) is that there is no need in bonding between two dielectric materials. One can simply fix the dielectric strips and then place the guiding layer (a dielectric plate). Mechanical pressure between the metal and dielectric plates is enough to make the structure operational. Also, only two dielectric regions are present, which simplifies the calculations and design and also possibly reduces losses. Different passive components (two directional couplers, a ring resonator) based on the inverted strip waveguide are proposed in [11].

In conclusion, two main approaches to the design of millimeter-wave integrated circuits on the base of dielectric waveguides exist nowadays. They use either non-radiating dielectric waveguides or strip dielectric waveguides (integrated optics). The non-radiating dielectric waveguide approach is developing more intensively due to the phenomenon of radiation suppression. However, the integrated optics technology is more suitable to implement the thin-film technology and incorporate three-terminal devices like transistors.

8.2 PASSIVE DEVICES

In this section we give an overview of passive (reciprocal and nonreciprocal) devices built on the base of millimeter-wave dielectric waveguides.

8.2.1 Whispering gallery resonator

When the basic properties of dielectric rod waveguides were studied in the 60's, it was shown that in addition to the usual modes propagating along the rod axis, in circular dielectric cylinders there are modes that propagate in the azimuthal direction, around the rod circumference [12]. These modes are referred to as *whispering gallery modes.* The fields of such modes are mainly concentrated near the circumference of the disk. Thus, it is possible to design resonant circuits using dielectric disks.

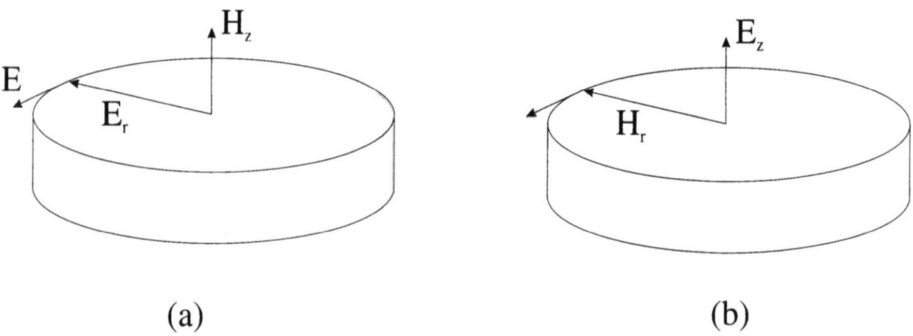

Figure 8.4. Electromagnetic fields of WGE (a) and WGH (b) modes.

The whispering gallery modes are usually classified as WGE_{nml} or WGH_{nml}, where the indices n, m, and l denote the numbers of azimuthal, radial and axial field variation periods, respectively [13,14], see Figure 8.4. The typical whispering gallery resonator is a dielectric disk supported in the middle, because the waves propagate mainly near the edge of the disk. Radiation losses of whispering modes can be negligible provided that the diameter of the disk is large enough. The quality factor of these resonators is limited by the dielectric losses and can be very high [13]. For example, Sapphire resonators at cryogenic temperatures about 77 K can have the quality factor as high as 30 million at 8 GHz [15]. At lower temperatures even higher quality factors are possible [16], since the loss tangent of Sapphire decreases with decreasing temperature: at 10 GHz and the temperature of 4 K the loss tangent of $7 \cdot 10^{-10}$ has been reported, which corresponds to the quality factor exceeding 10^9.

Such high quality factors allow designing stable and low-noise oscillators in different frequency bands with characteristics better than that of the conventional cavity resonators [15, 17–20]. Other applications of the whispering gallery resonators can be in band stop filters [14, 21], in measurements of the dielectric properties of materials [22, 23], when the resonant frequencies and quality factors of different modes are measured, and in power combining [14]. The advantage of this method is that it is not necessary to have a disk with a large diameter, for W frequency band the diameter about 15 mm is enough [22].

8.2.2 Directional couplers

One of the known approaches for designing millimeter-wave directional couplers is straightforward: the aperture coupling [24]. In this device, two dielectric waveguides are separated by a metal plate with a hole, as shown in Figure 8.5.

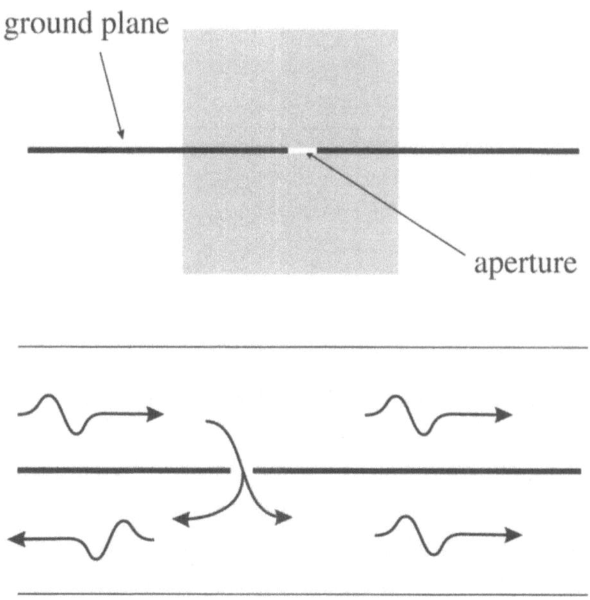

Figure 8.5. Aperture coupling configuration.

The second way is similar to that used in the microstrip line couplers. Such a directional coupler consists of two closely positioned transmission lines (Figure 8.6). In [25], a directional coupler of this type with $30 - 40$ dB isolation is described.

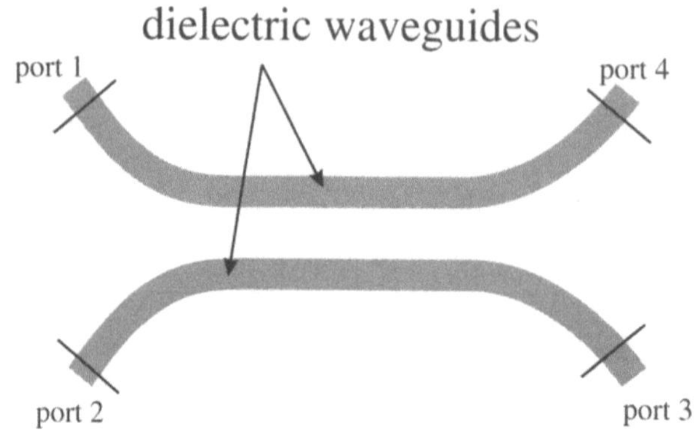

Figure 8.6. A directional coupler based on dielectric waveguides.

In the third approach, the second waveguide is connected to the first one like a probe. The end of the second dielectric waveguide is close to the surface of the first waveguide (Figure 8.7), see [26].

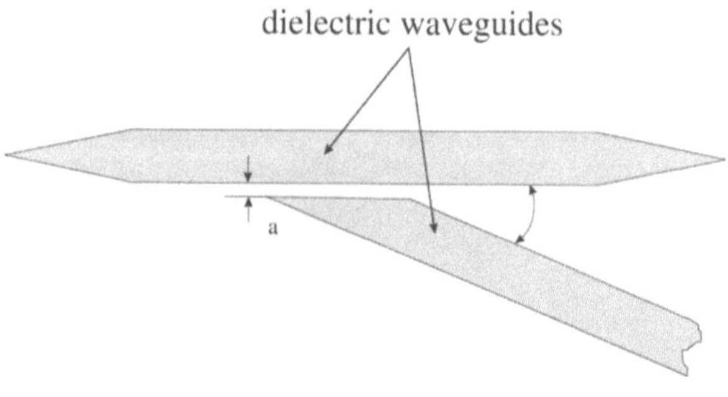

Figure 8.7. A probe-like directional coupler structure [26].

The coupling characteristics can be changed by changing the gap Δa and the angle θ. Directional couplers of this kind have more stable characteristics within the working frequency band, than the previous ones. Similar principles are described also in [27] for such directional couplers and for Y-junctions based on the same principle. Their application in modulator and switching elements is described in [28].

The next approach to designing directional couplers uses quasi-optical ideas [29, 30]. The principal structure is shown in Figure 8.8. This coupler consists of four dielectric waveguides and a reflecting di-

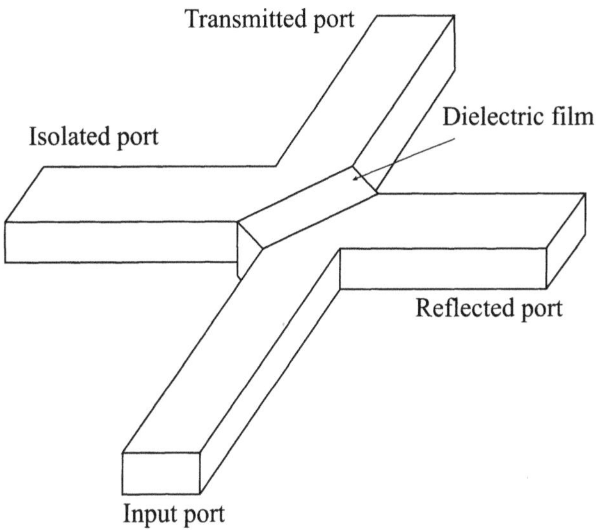

Figure 8.8. A quasi-optical directional coupler.

electric film, the permittivity and thickness of which determines the coupling coefficient and the operational frequency band. These parameters have to be chosen from the point of view of the minimal variations of the coupling coefficient vs. frequency. Experiments have shown its small variations with the frequency and the isolation changing from 20 dB at 26 GHz to more than 30 dB at 40 GHz. It can be thought that a stronger field concentration in the dielectric waveguide results in a better isolation. Especially, this effect has been observed when the dimensions of the dielectric rod waveguides were increased. Such directional coupler was used in so-called multistate reflectometer [31].

8.2.3 Phase shifters and attenuators

The general approach to phase shifting is to alter the propagation constant of the propagating wave along a fixed interval of the transmission line (a dielectric waveguide in our case). One of the straightforward methods to alter the propagation constant is to introduce perturbations near the surface of an open dielectric waveguide [31]. An example is shown in Figure 8.9. Changing the distance d between the surface of the dielectric waveguide and the metal plate one can change the wavelengths in the waveguide and thus control the phase shift.

Instead of introducing a metal plate one can place a semiconductor or

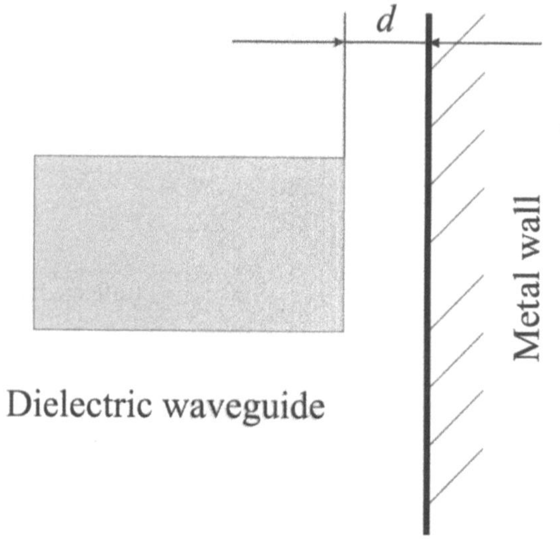

Figure 8.9. A mechanically controlled phase shifter.

a semiconductor p-i-n structure on the dielectric waveguide surface [32], which can be electrically controlled (Figure 8.10). The main disadvantages of such devices are mismatching and large insertion losses, e.g., losses about 6 dB have been reported in [32]. High losses are mainly due to contacts metallization, which is necessary to control the p-i-n junction.

The third way to produce a conducting layer is the optical injection of plasma by illuminating the waveguide with light radiation above the bandgap. The schematic structure is shown in Figure 8.11. It consists of an open dielectric waveguide with tapered ends to improve matching with standard metal waveguides, and a light source. It could be a laser located near the dielectric waveguide, or an optical transmission line. For example, in [34] a system producing 30 ps pulses at 0.53 μm wavelength with up to 50 μJ of energy was used. When the surface of the dielectric waveguide was illuminated with such a pulse, a 0.5 μm conducting layer appeared, changing the propagation constant of the dielectric waveguide and the signal phase shift.

Attenuation can be achieved applying radiation with a longer wavelength, e.g., in [34] 1.06 μm instead of 0.53 μm. In this case bulk conductivity is excited and thus the propagation losses dramatically increase.

The optical control principle gives the following main advantages. A very fast response (limited by the life time of charge carriers) is possible.

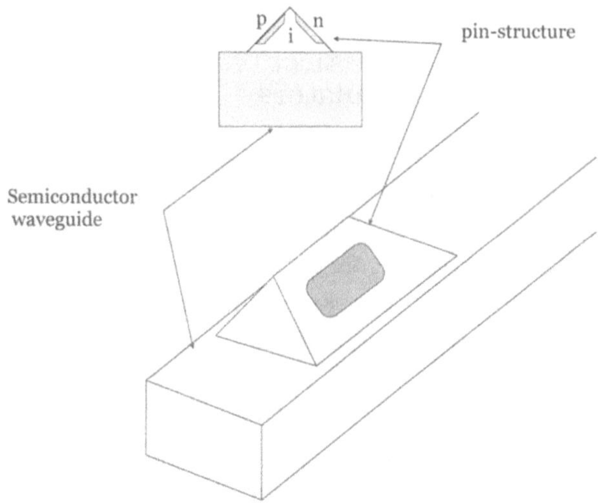

Figure 8.10. An electrically controlled phase shifter.

By a proper choice of the wavelength, the optical pulse width, and its energy, one can obtain the desired plasma density without damaging the material and thus obtain phase shifting and (or) attenuation. Very short millimeter-wave pulses can be generated this way.

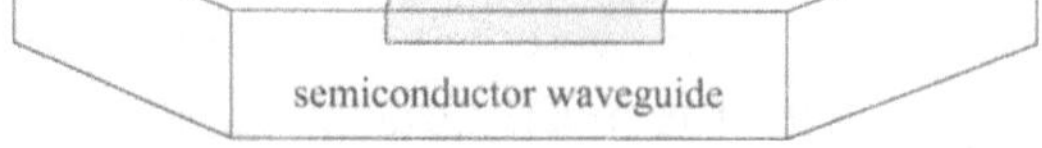

Figure 8.11. An optically controlled phase shifter.

Also, one can use magnetized ferrite waveguides to change the propagation constant, see, e.g., Figure 4.4. This device consists of a ferrite waveguiding structure and a magnetizing coil. Ferrite waveguide structures can be magnetized both longitudinally and transversally. However, the longitudinal magnetization seems to be more appropriate because of a smaller demagnetizing factor and thus weaker magnetic fields required. At millimeter-waves the performance of such systems is limited by commonly high loss tangents of ferrite materials and by the hysteresis of magnetization processes.

8.2.4 Isolators and circulators

Examples of different ferrite devices can be found e.g., in [33]. Isolators can be based on the Faraday rotation principle (Section 4.2, see also e.g. [33]), field displacement (see Section 4.3.1 and [35, 36]), and nonreciprocal power coupling between two transmission lines [37–40]. Circulators can be built on the same principles. Coupled mode analysis can be found e.g. in [41] with showing a possibility to design isolators with a high (more than 30 dB) isolation, but in a narrow frequency band. An isolator based on the nonreciprocal phase shift principle is proposed e.g. in [42]. It consists of a main dielectric waveguide and a ferrite pillbox working like a whispering gallery resonator.

8.3 ACTIVE DEVICES

Electromagnetic wave amplification and generation is one of the key problems in the design and technology of millimeter-wave devices and systems. It is convincingly shown by many authors (see e.g. [43]) that the output power of discrete semiconductor devices decreases with the frequency growth as approximately $1/f$, and drastically drops at frequencies above 100 GHz. Despite recent achievements in the semiconductor device technology, due to which the cut-off frequency of HEMTs, IMPATT and Gunn diodes is now near or above 300 GHz, and electromagnetic wave generation with resonant tunneling diode (RTD) was demonstrated at frequencies up to 700 GHz, the output power of these devices is extremely low. To overcome these fundamental limitations on the output power, which is mostly due to small sizes of discrete devices, the performance of active components should be based on the principles of distributed (in space) interaction of the electromagnetic wave with active media. Such principles are: i) power combining in multi-element grids and arrays; ii) traveling wave amplification in the waveguide, partly or completely filled with an active media, which could be a semiconductor in the state of bulk negative resistance; iii) the same traveling wave amplification as in ii) but due to an interaction of delayed electromagnetic wave with drifting charge flow (an analogue to the vacuum traveling wave tube).

Many authors studied power combining (see, e.g., [44]) both in the waveguide and quasi-optical configurations. It was shown that it is effective for the number of oscillators below 10 and at frequencies up to 20 GHz, but imposes poor efficiency in the millimeter-wave frequency region. Therefore, alternative principles like the traveling wave amplification, mentioned above, are of essential interest. Note, that traveling wave amplification should be more effective at millimeter-waves than at microwaves, particularly due to the fact that the length of the electromagnetic waves could be short enough for its effective interaction with active media at the propagation length in the waveguide. Usually the propagation length is limited by thermal heating, because all the known mechanisms of bulk negative resistance in semiconductors, which could be used for active waveguide performance, need high electric fields ($> 10^3$ V/cm) and current densities ($> 10^3$ A/cm^2) [45].

Amplification of slow traveling waves in monolithic solid-state structures that is due to the interaction with drifting electron flux such as proposed in [46] does not need bulk negative resistance and can be achieved at lower electric fields, than in the waveguide on a bulk negative resistance semiconductor layer. However, the problem is to provide a delay of electromagnetic waves with the delay factor of about 10^3 to make the phase velocity comparable with the electron drift velocity, which saturates in all known semiconductors at the value of about 10^7 cm/s [47]. Now both of the noted types of active waveguides have been theoretically and experimentally studied, but a further improvement in the theory and in the experimental performance appears to be necessary, as it will be shown below.

8.3.1 Theory of electromagnetic wave propagation in bulk negative resistance media

There are several phenomena in semiconductors that result in bulk negative resistance: negative effective mass, hot electron intervalley scattering, and some other effects [47]. The most promising mechanism of the bulk negative resistance for active waveguide performance is up to date the bulk negative differential resistance that arises in A^3B^5 semiconductors due to intervalley scattering of hot electrons in high electric fields. The advantage of this mechanism for active waveguide performance unlike to negative differential resistance in p-n junctions or quantum-well structures is that all the semiconductor volume that fills the waveguide structure could be active. The problem is, however, that the semiconductor contains free charges (electrons), so the process of electromagnetic wave propagation under bulk negative resistance conditions is strongly connected with excitation of space charge waves. The electromagnetic theory of such waveguides is considerably different in comparison with the theory of waves in dielectrics and dielectric waveguides, where the conduction current can be neglected, as well as with the theory of waveguides on semiconductors with a positive conductivity. In the last case, semiconductor inserts some losses, but no space charge wave can be excited.

An advanced theory of electromagnetic wave propagation in microstrip waveguides, which takes into account possible space charge wave excitation, was developed by some authors (see e.g., [45]) for the case of a semiconductor with an N-type current-voltage characteristic, which implies negative differential resistance in a certain interval of the bias voltage, where a semiconductor layer can be considered in the small signal approximation as a bulk negative resistance medium.

To simplify the mathematical modeling, the dependence of the current density on the electric field in the semiconductor is approximated by a step-wise linear function with three regions (Figure 8.12). This, in particular, quite adequately describes the $j - E$ curve for n-GaAs.

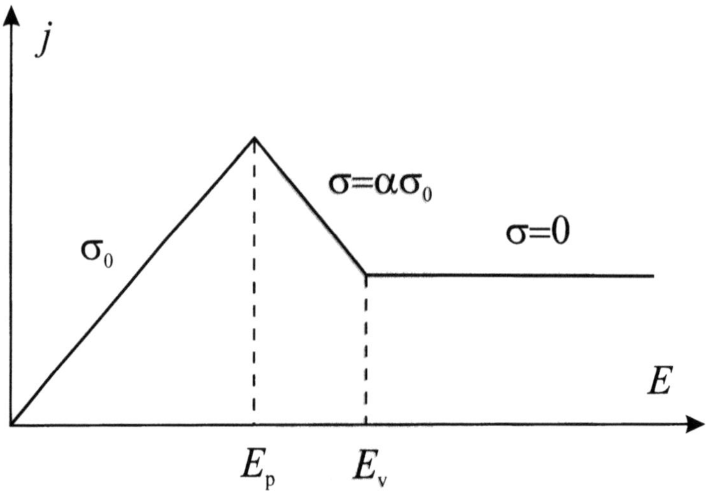

Figure 8.12. A step-wise linear approximation of the $j - E$ curve for a semiconductor. σ is the differential conductivity.

The microstrip waveguide (Figure 8.13) contains a semiconductor medium between two metallic plates, which are at the same time the ohmic contacts. When a bias voltage sufficient for creating a bulk negative resistance is applied and current instabilities (the Gunn effect) are suppressed [47], the steady-state field distribution becomes non-uniform, and at least in some regions of the semiconductor the field E can be larger than E_p, so that $\frac{dj}{dE} < 0$. Here, E_p corresponds to the peak value of the current density, see Figure 8.12. It is possible, if the electron concentration and the distance between the ohmic contacts are chosen accordingly to the criterium $\rho a < (1 \ldots 2) \cdot 10^{11}$ cm^{-2} (the product of the electron concentration and the distance) [47]. Thus, in typical cases the waveguide can be modeled as a multilayer structure with different conductivities of the layers (Figure 8.13).

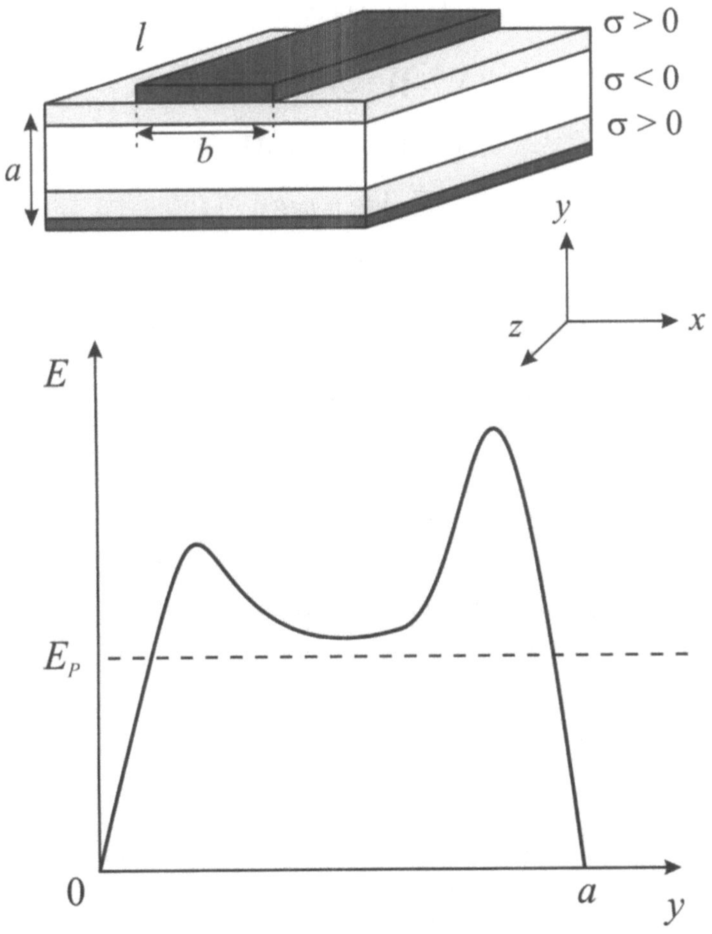

Figure 8.13. Microstrip waveguide schematic and a typical steady-state electric field distribution.

In order to find an analytical solution of the Maxwell equations in the small-signal approximation, one can substitute into the field equations the differential conductivity $\sigma = \frac{dj}{dE}$. It is anisotropic, as is obvious from Figure 8.13: $\sigma_x = \sigma_z = \sigma_0$, but σ_y depends on the applied bias electric field as shown in Figure 8.12. In the active layer the conductivity is considered as a uniaxial tensor

$$\hat{\sigma} = \begin{pmatrix} \sigma_0 & 0 & 0 \\ 0 & \alpha\sigma_0 & 0 \\ 0 & 0 & \sigma_0 \end{pmatrix} \tag{8.3}$$

where $\alpha = -\frac{1}{3}$ for n-GaAs.

For time-harmonic signals the Maxwell equations read:

$$\nabla \times \mathbf{H} = j\omega\varepsilon\mathbf{E} + \mathbf{j} \tag{8.4}$$

$$\nabla \times \mathbf{E} = -j\omega\mu_0\mathbf{H} \tag{8.5}$$

$$\nabla \cdot \mathbf{j} = -j\omega\rho \tag{8.6}$$

$$\nabla \cdot \mathbf{E} = \frac{\rho}{\varepsilon} \tag{8.7}$$

and they should be augmented by the equation for the current density:

$$\mathbf{j} + \mathbf{j}_0 = (\rho_0 + \rho)(\mathbf{v}_0 + \mathbf{v}) - D\nabla\rho \tag{8.8}$$

Here, ε is the dielectric constant in the absence of conduction current, μ_0 is the magnetic constant of the semiconductor, $\mathbf{j}_0$ and $\mathbf{j}$ are the DC and AC parts of the current density, ρ_0, ρ, $\mathbf{v}_0$, and $\mathbf{v}$ are the corresponding values of the space charge density and electron drift velocity, and D is the electron diffusion coefficient. The system of equations (8.4–8.8) can be transformed into the following:

$$\begin{gathered} \nabla^2\mathbf{E} + k^2\mathbf{E} = j\omega\mu_0\mathbf{j} + \tfrac{1}{\varepsilon}\nabla\rho \\ \nabla \cdot \mathbf{j} = -j\omega\rho \\ \mathbf{j} = \hat{\sigma} \cdot \mathbf{E} + \rho\mathbf{v}_0 - D\nabla\rho \end{gathered} \tag{8.9}$$

where $k = \omega\sqrt{\varepsilon\mu_0}$ is the propagation constant in the infinite medium in the absence of conduction current.

This system is too complicated to allow for an analytical solution, but it can be simplified as it was proposed in [48, 49], if one takes into account that the total field can be considered as a combination of two types of propagating waves: a "slow" space charge wave that propagates with the drift velocity of the electrons ($v_0 \approx 10^5$ m/s), and a "fast" electromagnetic wave that has the velocity close to the velocity of light.

With these assumptions, we can write:

$$\mathbf{E} = \mathbf{E}_{\text{fast}} + \mathbf{E}_{\text{slow}} \tag{8.10}$$

where $\mathbf{E}_{\text{fast}}$ is the electric field of the fast component, and $\mathbf{E}_{\text{slow}}$ is the electric field of the space charge wave. The total field $\mathbf{E}$ and the total current $\mathbf{j}$ propagate as a wave with the propagation constant k_{za}:

$$\mathbf{E}, \mathbf{j} \sim \exp(-jk_{za}z) \tag{8.11}$$

The total current can be written as

$$\mathbf{j} = \hat{\sigma} \cdot \mathbf{E} + \mathbf{j}_{\text{dr}} + \mathbf{j}_{\text{dif}} \tag{8.12}$$

Here, $\mathbf{j}_{\text{dr}}$ and $\mathbf{j}_{\text{dif}}$ are the drift and diffusion components of the space charge wave current. With these definitions the system of equations (8.9) can be transferred to:

$$\nabla_t^2 \mathbf{E} - k_{za}^2 \mathbf{E} + k^2 \mathbf{E} = j\omega\mu_0 \mathbf{j} + \frac{1}{\varepsilon} \nabla \rho \tag{8.13}$$

($\nabla_t = \nabla - \frac{\partial}{\partial z} \mathbf{z}_0$ is the transverse part of the gradient operator).

To solve this equation, a method similar to the perturbation analysis was developed in [48]. In that method, the parameters of slow space charge waves are first solved in the quasi-static approximation. Next, the propagation constant along the active waveguide k_{za} is found using the known transverse propagation constant of space charge waves.

For typical GaAs structures with a stabilized high-field distribution the expected gain was estimated in [48] (a few dB/mm at frequencies from 5 to 40 GHz).

In the fin-line geometry case (Figure 8.14) the mathematical formulation is more complicated, so special methods are needed to solve the corresponding equations. In [50], the method of autonomous blocks was applied. A theoretical evaluation shows a possibility of amplification of the traveling wave, as it is shown in Figure 8.15. This result and details of the modified electromagnetic theory are described in [45] and original papers, referenced therein. Theoretical considerations of active waveguides are also given in monograph [51].

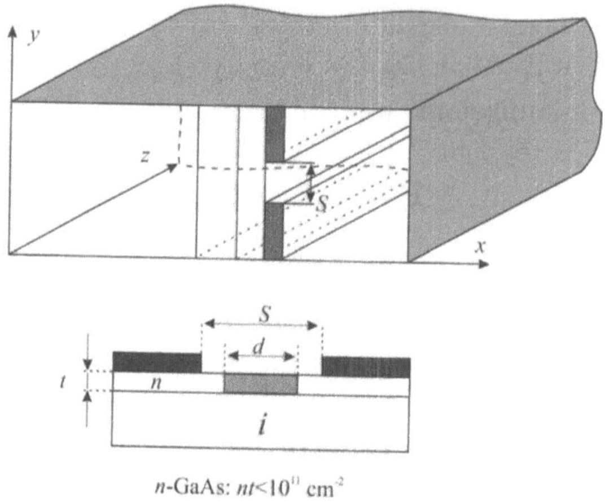

Figure 8.14. Schematic of a fin-line distributed amplifier.

8.3.2 Experimental observations of millimeter-wave amplification with active waveguides

The first experiments on electromagnetic wave amplification due to negative resistance in stabilized GaAs structures were performed in coplanar microstrip waveguides [52]. A high bias electric field was applied to metal thin-film strips, which were at the same time the ohmic contacts to an n-GaAs epitaxial layer on a semi-insulating GaAs substrate. The advantage of the planar geometry in this case, similar to the ordinary fin-lines, is that due to the influence of the dielectric substrate that suppresses the high-field domain excitation, there is a stabilization of the high field distribution. The thickness of the active n-layer must be small enough to satisfy the stability criterium $\rho t < (1 \ldots 2) \cdot 10^{11}$ cm^{-2}. The distance between the metal strips is not limited by this condition, and may be as long as 100 μm and more. This allows for a sufficient heat-sink when a high bias electric field is applied. However, this type of amplifiers has not developed into practical applications because of a high noise level and high insertion losses that limited the operation frequency at the level of approximately 15 GHz, so they yield in the competition with conventional transistor amplifiers.

Experiments with an active fin-line on GaAs were performed in [53]. An epitaxial GaAs layer on a semi-insulating GaAs substrate with $\rho = 8 \cdot 10^{13}$ cm^{-3} and the thickness $t = 30$ μm was stable at the applied bias electric field. The distance between the ohmic contacts was

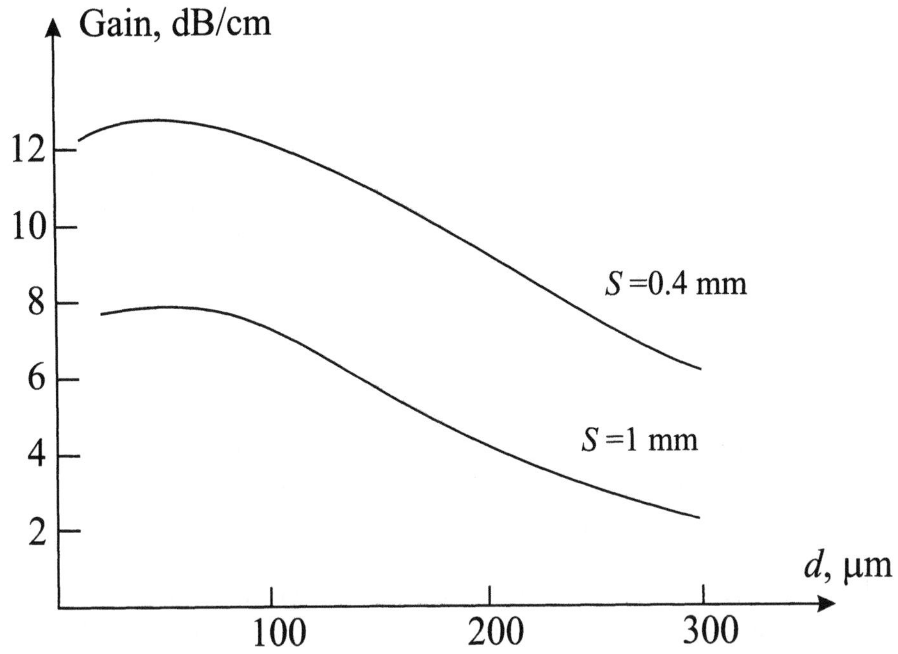

Figure 8.15. Electromagnetic wave gain in a fin-line waveguide with a negative resistance semiconductor.

$S = 1$ mm. The high-field steady-state domain was near the center, so the geometry was corresponding to the theoretical model illustrated in Figure 8.14. The compensation of losses ≈ 10 dB was demonstrated at the frequency 40 GHz, when 160 V bias voltage was applied in the pulse regime. Further advances of this amplifier were not successful because of too heavy conditions of heat-sink and high insertion losses in the metal strips.

Attempts to realize traveling wave amplification in structures like a non-symmetrical strip line, as shown in Figure 8.13, were also made [54]. They demonstrated an amplification and in some cases signal generation, if the waveguide length l was close to $\frac{\lambda}{2}$, so the waveguide section operated as a resonator. The problem is common for all waveguide constructions: bulk negative resistance is realized in GaAs and other semiconductors at high electric fields ($> 10^3$ V/cm) and current densities ($> 10^3$ A/cm^2). Therefore, the thermal heating is too intensive for amplification in the continuous regime that is of the main practical interest. Future prospects of such amplifiers are possible, if bulk negative resistance could be realized at lower current densities, e.g., about

10^2 A/cm^2. Some evidence of this opportunity with resonance-tunneling structures appeared recently [55], so a return to active waveguides on semiconductors with bulk negative resistance is possible in the future.

If amplification of "fast" electromagnetic waves in waveguides containing negative resistance media looks problematic, nonlinear effects, like frequency multiplication, based on the presence of conductance or capacitance dependent on the electric field, seems to be promising at millimeter-wave frequencies. The results of experiments on frequency multiplication in a waveguide containing Schottky barrier or heterobarrier structures [56] give certain confirmations to this.

8.3.3 Slow electromagnetic wave amplification with drifting electrons in semiconductor waveguide structures

The idea of traveling wave amplification in a semiconductor with a drifting electron flow analogous to vacuum traveling wave tubes is well known since the work of L. Solimar and E. Ash [57]. Experimental demonstrations were performed at microwave frequencies, though it was clear that only at millimeter waves this mechanism is of practical interest due to short electromagnetic wavelengths. The problem was to find a type of the delay structure which could provide a sufficient decrease of the phase velocity without considerable dispersion at millimeter-wave frequencies. Such appropriate structure – periodically corrugated image waveguide – was proposed and theoretically analyzed in [46]. Following this idea, GaAs epitaxial structures were used as image-type waveguides in [58] (Figure 8.16). In some cases the upper i-layer was made of AlGaAs.

Thin epitaxial n- and i-layers were formed in the growth process by MOCVD. A delay of the electromagnetic wave was provided by a corrugated surface, which was manufactured by holographic lithography combined with dry plasma or photochemical etching. The delay structures had a period of 0.5-1.0 mm with 0.1 mm groove depth. Electron concentration in the n-type layer was $10^{15} - 10^{16}$ cm^{-3}.

In the general case, when an electromagnetic wave propagates in a periodical structure, its field can be represented as a series of spatial harmonics, as determined by the Floquet theorem:

$$E_z\left(y,z\right) = \sum_{m=-\infty}^{\infty} a_m(y)\exp\left(-jk_{zm}z\right) \tag{8.14}$$

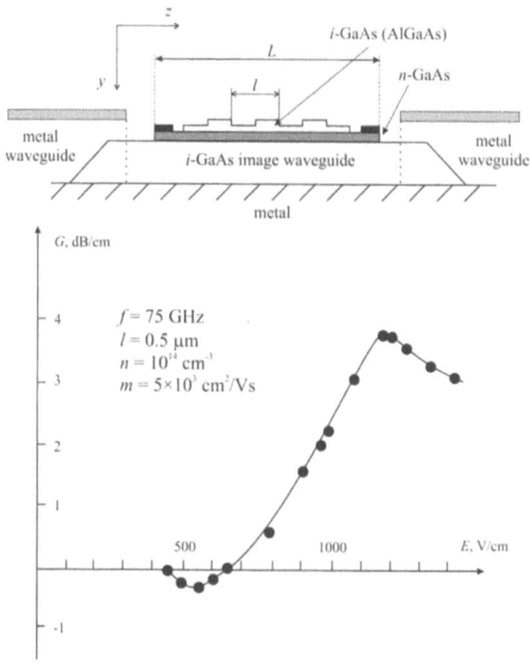

Figure 8.16. Scheme of the experiment and experimental results of slow wave amplification [58].

where $k_{zm} = k_0 + \frac{2\pi}{l}m$, l is the period of the grating, and k_0 is the propagation constant of the unperturbed waveguide (see Section 1.6). Gain occurs when the drift velocity of the electrons becomes slightly higher than the phase velocity of the electromagnetic wave component. The condition for the first space harmonic is given by:

$$V_{\rm d} > V_{\rm ph} = \frac{\omega}{k_0 + \frac{2\pi}{l}} \tag{8.15}$$

where $V_{\rm d}$ is the carrier drift velocity, and ω is the angular frequency. For the expected case with $k_0 \ll \frac{2\pi}{l}$, the amplification condition is $V_{\rm d} > fl$, where f is the signal frequency. The electromagnetic energy of space harmonics is concentrated within the layer about $\frac{l}{2\pi}$ thick, which is feasible in A^3B^5 heterostructures, like the structures used in [58]. Evaluations made in [46] show that this mechanism of traveling wave amplification is effective at frequencies about 100 GHz, where it is of essential practical interest.

Note, that the value of the gain threshold field (650 V/cm in [58]) is considerably lower than the electric field in majority of millimeter-wave semiconductor devices and active waveguides based on bulk negative resistance. Taking into account good opportunities of matching of dielectric rod waveguides with metal waveguides (see Chapter 7 and [59]), it seems to be possible to realize true input-output amplification (in [58] only "electronic" gain was observed). Therefore, the traveling wave type amplification appears to be promising for the design of low noise millimeter-wave devices and integrated circuits.

Heterostructures AlGaAs/GaAs and AlGaN/GaN with two-dimensional electron gas layers seem to be prospective as materials for traveling wave amplifiers. In this case the amplification of a slow electromagnetic wave by a drifting electron flow can be much more effective, than it was achieved in the experiment described in [58], due to high mobility of the electrons. The presence of potential barriers in AlGaAs/GaAs and AlGaN/GaN junctions could also be used for nonlinear wave transformations during electromagnetic wave propagation.

8.4 DIELECTRIC WAVEGUIDE ANTENNAS

As we have already seen, dielectric materials can have rather attractive properties like lower loss, cost, much larger tolerances for dimensions of devices made of such materials, as compared to metal devices. Therefore, also millimeter-wave antennas made of dielectric materials should be considered.

8.4.1 Classification

A schematic classification of dielectric antennas is shown in Figure 8.17. Dielectric antennas can be subdivided into antennas based on dielectric waveguides (traveling wave antennas) and lens antennas. The traveling wave antennas can be classified depending on the main direction of the beam propagation: broad-side and end-fire antennas. An example of an end-fire antenna is a tapered dielectric rod [60–62]. An example of a broadside antenna is the so-called leaky-wave antenna (dielectric rod with discontinuities on its surface) [63–66]. A review of different antennas can be found e.g. in [67]. Here, the attention will be paid to antennas based on the dielectric waveguide.

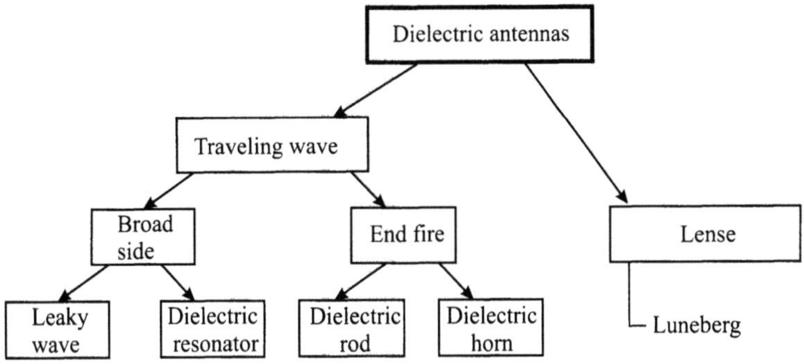

Figure 8.17. Main types of dielectric antennas.

8.4.2 Dielectric rod antennas

The common configuration of a dielectric rod antenna is shown in Figure 8.18.

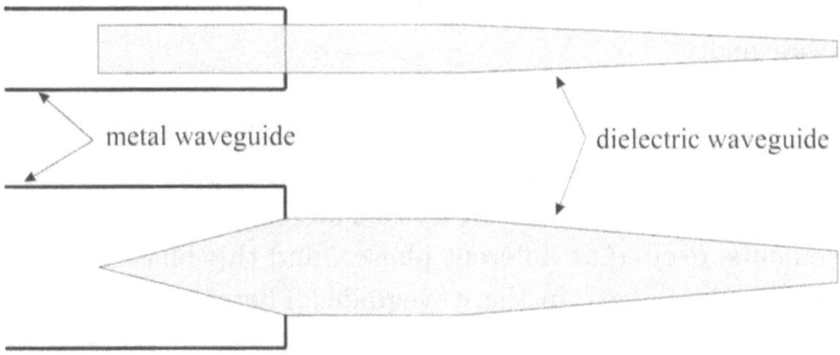

Figure 8.18. Common configuration of a dielectric rod antenna.

The antenna consists of a transition from a metal to a dielectric waveguide, followed by a section of the dielectric waveguide itself, which is tapered at the end to provide radiation from the whole tapering length and decrease the reflection coefficient from the end. Dielectric rod antennas have usually a relatively wide beamwidth and a moderate side lobe level. For comparison [68], a 40 mm long dielectric rod antenna with 1.6×3.2 mm^2 cross-section has the -15.5 dB side lobe level with 3 dB beamwidth of 23.5°, while a 40 mm long metal rectangular horn has -11.5 dB side lobe level and 14.2° 3 dB beamwidth.

8.4.3 Leaky-wave antennas

The dielectric waveguide is an open transmission line, therefore at any discontinuity at its surface radiation in the surrounding media occurs. Waves traveling along the waveguide and radiating into space from the surface of the dielectric waveguide are called leaky waves. Traveling-wave antennas that use leaky wave modes in dielectric waveguides can be designed.

Usually, a leaky-wave antenna is a dielectric waveguide of an arbitrary cross section with periodical discontinuities on its surface, such as notches or metal strips (Figure 8.19).

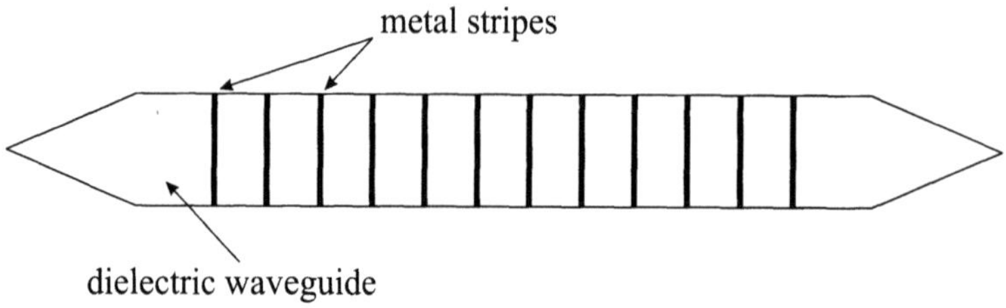

Figure 8.19. Common configuration of a leaky-wave antenna based on a dielectric waveguide.

As one dimension (the length) of a leaky-wave antenna can be quite large, it is possible to obtain a narrow radiation pattern in one plane. Moreover, the discontinuities can be considered as arrays of small radiating elements, excited at different phases, and this phase difference depends on the wavelength in the waveguide. Therefore, frequency beam scanning is possible, because the propagation factor in the waveguide depends on the operational frequency.

The main beam direction can be estimated using the following formula [69]:

$$\theta_m = \arcsin\left(\frac{\lambda_0}{\lambda_g} + \frac{\lambda_0}{d}m\right), \qquad \left|\frac{\lambda_0}{\lambda_g} + \frac{\lambda_0}{d}m\right| < 1 \tag{8.16}$$

where θ_m is the m-th beam angle from the broadside (the normal), λ_0 is the wavelength in free space, λ_g is the wavelength in the dielectric waveguide, and m is the space harmonic number. This antenna [69] is proposed for frequencies up to 100 GHz, and it has a rather wide scan range between 68° and 7° at VSWR less than 1.4, and the half-power beamwidth of about 4° in one plane.

The main disadvantage of such antennas is that not all the power propagating along the dielectric waveguide is radiated. The rest of the power causes a peak in the end fire direction, and to prevent this effect the power reaching the waveguide end should be absorbed. Therefore, the antenna efficiency is not very high, and its typical value is approximately 0.87.

There has been an interesting attempt to combine leaky wave and horn antennas [70]. The cross-section of such an antenna is shown in Figure 8.20. This antenna is proposed for operation in the W frequency band and has 3° half-power beamwidth with 26 dB of gain at the optimal flare angle of the horn, and a very low side lobe level.

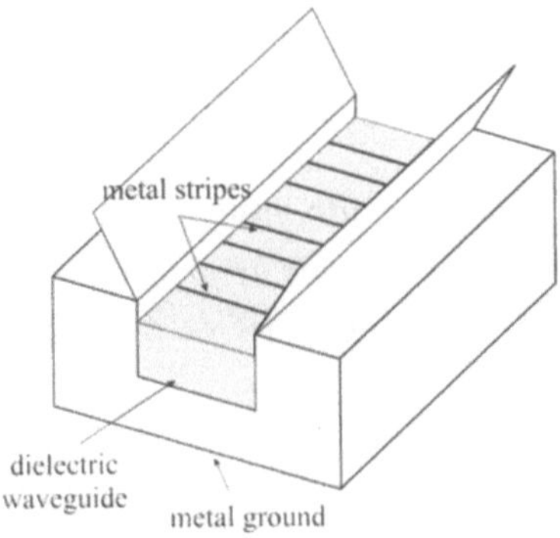

Figure 8.20. Horn leaky-wave antenna.

References

[1] Y. Tokumitsu, M. Ishizaki, M. Iwakuni, and T. Saito, 50-GHz IC components using alumina substrate, *IEEE Transactions on Microwave Theory and Techniques*, vol. MTT-31, no. 2, February 1983, pp. 121-128.

[2] T. Yoneyama, Millimeter-wave integrated circuits using nonradiative dielectric waveguide, *Electronics and Communications in Japan*, Part 2, vol. 74, no. 2, 1991, pp. 20-28.

[3] T. Yoneyama and S. Nishida, Nonradiative dielectric waveguide for millimeter-wave integrated circuits, *IEEE Transactions on Microwave Theory and Techniques*, vol. MTT-29, no. 11, November 1981, pp. 1188-1192.

[4] T. Yoneyama and S. Nishida, Experimental design of millimeter-wave nonradiative dielectric waveguide filters, *Electronics and Communications in Japan*, Part 1, vol. 69, no. 5, 1986, pp. 83-89.

[5] K. Wu, Hybrid integration technology of planar circuits and NRD-guide for cost-effective microwave and millimeter-wave applications, *IEEE Transactions on Microwave Theory and Techniques*, vol. 45, no. 6, June 1997, pp. 946-954.

[6] H. Kogelnik, An introduction to integrated optics, *IEEE Transactions on Microwave Theory and Techniques*, vol. MTT-23, no. 1, January 1975, pp. 2-16.

[7] M. Chrepta and H. Jacobs, Millimeter-wave integrated circuits, *Microwave J.*, vol. 17, no. 11, November 1974, pp. 45-47.

[8] R. Knox, Dielectric waveguide microwave integrated circuits – an overview, *IEEE Transactions on Microwave Theory and Techniques*, vol. MTT-24, no. 11, November 1976, pp. 806-814.

[9] W. McLevige, T. Itoh, and R. Mittra, New waveguide structures for millimeter-wave and optical integrated circuits, *IEEE Transactions on Microwave Theory and Techniques*, vol. MTT-23, no. 10, October 1975, pp. 788-794.

[10] T. Itoh, Inverted strip dielectric waveguide for millimeter-wave integrated circuits, *IEEE Transactions on Microwave Theory and Techniques*, vol. MTT-24, no. 11, November 1976, pp. 821-827.

[11] R. Rudokas and T. Itoh, Passive millimeter-wave IC components made of inverted strip dielectric waveguides, *IEEE Transactions on Microwave Theory and Techniques*, vol. MTT-24, no. 12, December 1976, pp. 978-981.

[12] J.R. Wait, Electromagnetic whispering gallery modes in a dielectric rod, *Radio Science*, vol. 2, no. 2, 1967, pp. 1005-1017.

[13] M.E. Tobar, J.G. Hartnett, E.N. Ivanov, P. Blondy, and D. Cros, Whispering gallery method of measuring complex permittivity in highly anisotropic materials: discovery of a new type of mode in anisotropic dielectric resonators, *IEEE Transactions on Instrumentation and Measurement*, vol. 50, no. 2, April 2001, pp. 522-525.

[14] D. Cros and P. Guillon, Whispering gallery dielectric resonator modes for W-band devices, *IEEE Transactions on Microwave Theory and Techniques*, vol. 38, no. 11, November 1990, pp. 1667-1674.

[15] G.J. Dick, D.G. Santiago, and R.T. Wang, Temperature-compensated sapphire resonator for ultra-stable oscillator capability at temperatures above 77K, *IEEE Transactions on Ultrasonics, Ferroelectrics and Frequency Control*, vol. 42, no. 5, September 1995, pp. 812-819.

[16] V.B. Braginsky, V.S. Ilchenko, and Kh.S. Bagdassarov, Experimental observation of fundamental microwave absorption in high-quality dielectric crystals, *Physics Letters A*, vol. 120, no. 6, 2 March 1987, pp. 300-305.

[17] M.E. Tobar, E.N. Ivanov, P. Blondy, D. Cros, and P. Guillon, High-Q whispering gallery traveling wave resonators for oscillator frequency stabilization, *IEEE Transactions on Ultrasonics, Ferroelectrics and Frequency Control*, vol. 47, no. 2, March 2000, pp. 421-425.

[18] M.E. Tobar and D.G. Blair, Phase noise analysis of the sapphire loaded superconducting niobium cavity oscillator, *IEEE Transactions on Microwave Theory and Techniques*, vol. 42, no. 2, February 1994, pp. 344-347.

[19] D. Cros, C. Tronche, P. Guillon, and B. Theron, W band whispering gallery dielectric resonator mode oscillator, *IEEE MTT-S International Microwave Symposium Digest*, vol. 3, 1991, pp. 929-932.

[20] S.L. Badnikar, N. Shanmugam, and V.R.K. Murthy, Microwave whispering gallery mode dielectric resonator oscillator, *Proceedings of the Joint Meeting of the European Frequency and Time Forum and the IEEE International Frequency Control Symposium*, 1999, vol. 2, pp. 597-600.

[21] X.H. Jiao, P. Guillon, L.A. Bermudez, and P. Auxemery, Whispering-gallery modes of dielectric structures: Applications to millimeter-wave bandstop filters, *IEEE Transactions on Microwave Theory and Techniques*, vol. MTT- 35, no. 12, December 1987, pp. 1169-1175.

[22] P. Blondy, D. Cros, P. Guillon, F. Balleras, and C. Massit, W band silicon dielectric resonator for semiconductor substrate characterization, *IEEE MTT-S International Microwave Symposium Digest*, vol. 3, 1998, pp. 1349- 1352.

[23] K. Derzakowski, A. Abramowicz, and J. Krupka, Whispering gallery resonator method for permittivity measurements, *Proc. of 13th International Conference on Microwaves, Radar and Wireless Communications, MIKON-2000,* vol. 2, 2000, pp. 425-428.

[24] I. Bahl and P. Bhartia, Aperture coupling between dielectric image limes, *IEEE Transactions on Microwave Theory and Techniques*, vol. MTT-29, September 1981, pp. 891-896.

[25] K. Solbach, The calculation and the measurement of the coupling properties of dielectric image lines of rectangular cross section, *IEEE Transactions on Microwave Theory and Techniques*, vol. MTT-27, January 1979, pp. 54-58.

[26] S.S.Gigoyan and B.A. Murmuzhev, Wide band couplers based on image dielectric waveguides, *Radiotekhnika*, no. 2, 1988, pp. 86-87 (in Russian).

[27] K. Ogusu, Experimental study of dielectric waveguide Y-junction for millimeter-wave integrated circuits, *IEEE Transactions on Microwave Theory and Techniques*, vol. MTT-33, no. 6, June 1985, pp. 506-509.

[28] A. Axelrod and M. Kisliuk, Experimental study of the W-band dielectric-guide Y-branch interferometer, *IEEE Transactions on Microwave Theory and Techniques*, vol. MTT-32, no. 1, January 1984, pp. 46-50.

[29] R. Collier and G. Hjipieris, A broad-band directional coupler for both dielectric and image guides, *IEEE Transactions on Microwave Theory and Techniques*, vol. MTT-33, no. 2, February 1985, pp. 161-163.

[30] R.D. Birch and R.J. Collier, A broadband image guide directional coupler, *Proceedings of 10th European Microwave Conference*, Warszawa, 8-11 September 1980, pp. 295-298.

[31] R.J. Collier and M.F. D'Souza, A multistate reflectometer in dielectric guide, *IEE Colloquium on What's New in Microwave Measurements*, 1990, pp. 10/1-10/10.

[32] H. Jacobs and M.M. Chrepta, Electronic phase shifter for millimeter-wave semiconductor dielectric integrated circuits, *IEEE Transactions on Microwave Theory and Techniques*, vol. MTT-22, no. 4, 1974, pp. 411-417.

[33] J. Helszajn, *Ferrite Phase Shifters and Control Devices*, McGraw-Hill book company, 1989.

[34] C.H. Lee, P.S. Mak, and A.P. DeFonzo, Optical control of millimeter-wave propagation in dielectric waveguides, *IEEE J. of Quantum Electronics*, vol. QE-16, no. 3, 1980, pp. 277-288.

[35] R.W. Babbitt and R.A. Stern, Millimeter wave ferrite devices, *IEEE Transactions on Magnetics*, vol. MAG-18, no. 6, November 1982, pp. 1592-1594.

[36] J.M. Owens, J.Y. Guo, W.A. Davis, and R.L. Carter, W-band ferrite-dielectric image-line field displacement isolators, *IEEE MTT-S Digest*, 1989, pp. 141-144.

[37] M. Tsutsumi and K. Kumagai, Dielectric slab waveguide isolator in the millimeter wave frequency, *IEEE Transactions on Magnetics*, vol. MAG-23, no. 5, September 1987, pp. 1739-3740.

[38] K. Tanaka, M. Tsutsumi, and N. Kumagai, Millimeter wave dielectric waveguide isolator, *Electronics and Communications in Japan*, Part2, vol. 71, no. 10, 1988, pp. 92-100.

[39] S.S. Gigoyan and B.A. Murmuzhev, Ferrite isolator for millimeter wave region based on image dielectric waveguide, *Radiotekhnika*, no. 4, 1986, pp. 84-85 (in Russian).

[40] A.A. Ahumyan, S.S. Gigoyan, B.A. Murmuzhev, and P.M. Martirosyan, Ferrite isolators for millimeter wave region based on image dielectric waveguide made of alumina, *Radiotekhnika*, no. 2, pp. 41-43 (in Russian).

[41] I. Awai and T. Itoh, Coupled-mode theory analysis of distributed nonreciprocal devices, *IEEE Transactions on Microwave Theory and Techniques*, vol. MTT-29, no. 10, October 1981.

[42] M. Muraguchi, K. Araki, and Y. Naito, A new type of isolator for millimeter-wave integrated circuits using a nonreciprocal traveling-wave resonator, *IEEE Transactions on Microwave Theory and Techniques*, vol. 30, no. 11, November 1982.

[43] V.E. Lybchenko, Fundamental limitations and prospects of semiconductor sevice application in millimetre wave radio systems, in *Physics and Technology of Millimetre Wave Components and Devices*, London and New York: Taylor and Francis, 2002, pp. 1-26.

[44] R.A. York and Z.B. Popoviĉ (Editors), *Active and Quasi-Optical Arrays for Solid-State Power Combining*, New York: Wiley, 1997.

[45] V.E. Lybchenko, Active MMW dielectric waveguides, in *Physics and Technology of Millimetre Wave Components and Devices*, London and New York: Taylor and Francis, 2002, pp. 43-54.

[46] A. Grover and A. Yariv, Monolithic solid-state travelling-wave amplifiers, *J. of Applied Physics*, vol. 45, 1976, pp. 2596-2600.

[47] M. Shur, *Physics of Semiconductor Devices*, Englewood Cliffs, N.J.: Prentice-Hall, 1990.

[48] V.E. Lyubchenko and G.S. Makeeva, Electrodynamical analysis of EMW propagation in a thin film semiconductor structure with negative differential conductivity, *Radiotekhnika i Elektronika*, vol. 28, no. 8, 1983, pp. 1633-1641 (in Russian).

[49] V.E. Lyubchenko and G.S. Makeeva, Electrodynamical analysis of the active fin-line, *Radiotekhnika i Elektronika*, vol. 28, no. 11, 1983, pp. 2102-2107 (in Russian).

[50] O.A. Golovanov, V.E. Lyubchenko, and G.S. Makeeva, Computer modeling of active fin-lines on GaAs with high-field domain, *Elektronnaja Tekhnika*, ser. 1 (Microwave Electronics), no. 12, 1978, pp. 8-10 (in Russian).

[51] A.A. Barybin, *Waves in Thin-Film Semiconductor Structures with Hot Electrons,* Moscow: Nauka, 1986 (in Russian).

[52] P. Fleming, The active medium propagation device, *Proceedings of the IEEE*, vol. 63, no. 8, 1975, pp. 1253-1254.

[53] V.I. Borisov, T.A. Briantseva, V.A. Korobkin, and V.E. Lyubchenko, Amplification of electromagnetic waves in the fin-line on GaAs, *Radiotekhnika i Elektronika*, no. 1, 1981, pp. 173-176 (in Russian).

[54] A.A. Barybin and V.M. Prigorovsky, *Elektronnaja Tekhnika*, ser. 1 (Microwave Electronics), no. 12, 1978, pp. 8-10 (in Russian).

[55] J.P.A. Van Der Vagt, Tunneling-based SRAM, *Proceedings of IEEE*, vol. 87, no. 4, 1999, pp. 571-596.

[56] J.R. Trope, D.P. Stenson, and R.E. Miles, Nonlinear transmission lines for millimeter-wave frequency amplifier application, *Terahertz Electronics Proceedings, IEEE 6-th International Conference*, 1998, pp. 54-57.

[57] L. Solymar and E. Ash, *International J. of Electronics*, no. 20, 1966, pp. 127-148.

[58] V.E. Lyubchenko and V.A. Martiakin, Travelling wave amplification in GaAs image waveguide at V band, *Electronics Letters*, vol 30, no. 11, 1994, pp. 869-870.

[59] D. Lioubtchenko, S. Dudorov, J. Mallat, J. Tuovinen, and A.V. Räisänen, Low-loss sapphire waveguides for 75-110 GHz frequency range, *IEEE Microwave and Wireless Components Letters*, vol. 11, no. 6, 2001, pp. 252-254.

[60] A.C. Studd, Towards a better dielectric rod antenna, *Proceedings of IEE Seventh International Conference on Antennas and Propagation*, 1991, pp. 117-120.

[61] Y. Shiau, Dielectric rod antennas for millimeter-wave integrated circuits, *IEEE Transactions on Microwave Theory and Techniques*, vol. MTT-24, no. 11, 1976, pp. 869-872.

[62] M. Aubrion, A. Larminat, and H. Baudrand, Design of a dual dielectric rod-antenna system, *IEEE Microwave and Guided Wave Letters*, vol. 3, no. 8, 1993, pp. 276-277.

[63] A. Basu and T. Itoh, Dielectric waveguide-based leaky-wave antenna at 412 GHz, *IEEE Transactions on Antennas and Propagation*, vol. 46, no. 11, 1998, pp. 1665-1673.

[64] K. Solbach, E-band leaky wave antenna using dielectric image line with etched radiating elements, *IEEE MTT-S International Microwave Symposium*, April 30-May 2, 1979, pp. 214-216.

[65] K.K. Narayanan, K. Vasudevan, and K.G. Nair, A dielectric rod leaky-wave antenna with a conducting ground plane, *Antennas and Propagation Society International Symposium*, 1988, vol. 1, pp. 366-369.

[66] S. Kobayashi, R. Lampe, R. Mittra, and S. Ray, Dielectric rod leaky-wave antennas for millimeter-wave applications, *IEEE Transactions on Antennas and Propagations*, vol. AP-29, no. 5, 1981, pp. 822-824.

[67] K. Solbach, in *Infrared and Millimeter Waves*, vol. 15, Orlando, FL: Academic Press, 1986, pp. 193-219.

[68] S. Kobayashi, R. Lampe, N. Deo, and R. Mittra, Dielectric antennas for millimeter-wave applications, *IEEE MTT-S International Microwave Symposium*, 1979, pp. 566-568.

[69] K. Klohn, R. Horn, H. Jacobs, and E. Freibergs, Silicon waveguide frequency scanning linear array antenna, *IEEE Transactions on Microwave Theory and Techniques*, vol. MTT-26, no. 10, 1978, pp. 764-773.

[70] T.N. Trinh, R. Mittra, and R.J. Paleta, Horn image-guide leaky-wave antenna, *IEEE Transactions on Microwave Theory and Techniques*, vol. MTT-29, December 1981, pp. 1310-1314.

Appendix A: Dyadics

In theoretical physics, linear operators in vector spaces are usually defined in terms of tensors and their corresponding matrices. Dyadics provide a convenient alternative in the three-dimensional space. It is possible to establish a one-to-one correspondence between tensor and dyadic spaces. The main advantage of the dyadic formalism is that dyadics are built up from vectors which often have a clear physical meaning. Using vector algebra, many powerful theorems of dyadic algebra can be established [1], and some of them have no simple equivalent in tensor algebra. Another advantage is that this formalism is coordinate independent.

Definitions and dyadic algebra

Dyad is a pair of vectors: **ab** (which is not the same as **ba**). Dyadic is a linear combination (introduced formally on this stage) of dyads: $\overline{\overline{A}} = \mathbf{ab} + \mathbf{cd} + \mathbf{ef} + \ldots$.

Multiplication by a scalar α is defined as

$$\alpha\overline{\overline{A}} = (\alpha\mathbf{a})\mathbf{b} = \mathbf{a}(\alpha\mathbf{b}) \tag{17}$$

Addition satisfies

$$\mathbf{ab} + \mathbf{cb} = (\mathbf{a} + \mathbf{c})\mathbf{b}, \qquad \mathbf{ab} + \mathbf{ac} = \mathbf{a}(\mathbf{b} + \mathbf{c}) \tag{18}$$

Scalar multiplication by a vector is defined by:

$$\overline{\overline{A}} \cdot \mathbf{x} = (\mathbf{ab} + \mathbf{cd} + \mathbf{ef} + \ldots) \cdot \mathbf{x} = \mathbf{a}(\mathbf{b} \cdot \mathbf{x}) + \mathbf{c}(\mathbf{d} \cdot \mathbf{x}) + \mathbf{e}(\mathbf{f} \cdot \mathbf{x}) + \ldots \tag{19}$$

$$\mathbf{x} \cdot \overline{\overline{A}} = (\mathbf{x} \cdot \mathbf{a})\mathbf{b} + (\mathbf{x} \cdot \mathbf{c})\mathbf{d} + (\mathbf{x} \cdot \mathbf{e})\mathbf{f} + \ldots \tag{20}$$

This operation defines a linear operator acting on vectors:

$$\mathbf{g} = \overline{\overline{A}} \cdot \mathbf{x} \tag{21}$$

Vector multiplication by a vector is defined through

$$\overline{\overline{B}} = \overline{\overline{A}}\times\mathbf{z} = (\mathbf{ab}+\mathbf{cd}+\mathbf{ef}+\ldots)\times\mathbf{z} = \mathbf{a}(\mathbf{b}\times\mathbf{z})+\mathbf{c}(\mathbf{d}\times\mathbf{z})+\mathbf{e}(\mathbf{f}\times\mathbf{z})+\ldots \tag{22}$$

$$\overline{\overline{C}} = \mathbf{z}\times\overline{\overline{A}} = \mathbf{z}\times(\mathbf{ab}+\mathbf{cd}+\mathbf{ef}+\ldots) = (\mathbf{z}\times\mathbf{a})\mathbf{b}+(\mathbf{z}\times\mathbf{c})\mathbf{d}+(\mathbf{z}\times\mathbf{e})\mathbf{f}+\ldots \tag{23}$$

Transposition simply changes the order of vectors in pairs:

$$\overline{\overline{A}}^T = (\mathbf{ab}+\mathbf{cd}+\mathbf{ef}+\ldots)^T = \mathbf{ba}+\mathbf{dc}+\mathbf{fe}+\ldots \tag{24}$$

In a given basis, every dyadic can be uniquely expressed in terms of pairs of basis vectors $\mathbf{e}_i$ $(i = 1, 2, 3)$:

$$\overline{\overline{A}} = A^{ij}\mathbf{e}_i\mathbf{e}_j = A^{11}\mathbf{e}_1\mathbf{e}_1 + A^{12}\mathbf{e}_1\mathbf{e}_2 + \cdots + A^{33}\mathbf{e}_3\mathbf{e}_3 \tag{25}$$

Nine coefficients A^{ij} form the matrix of the dyadic in this coordinate frame. It can be shown that coefficients A^{ij} transform as components of a second-rank tensor. This way we establish correspondence between tensors, matrices, and dyadics.

The unit dyadic. Symmetric and antisymmetric dyadics

By definition, the unit dyadic corresponds to the identity operator:

$$\overline{\overline{I}}\cdot\mathbf{a} = \mathbf{a} \tag{26}$$

for all vectors $\mathbf{a}$. For example, in Cartesian coordinates

$$\overline{\overline{I}} = \mathbf{x}_0\mathbf{x}_0 + \mathbf{y}_0\mathbf{y}_0 + \mathbf{z}_0\mathbf{z}_0 \tag{27}$$

From this representation it is obvious that the unit dyadic is symmetric: $\overline{\overline{I}}^T = \overline{\overline{I}}$, thus, for any $\mathbf{a}$ we have $\mathbf{a}\cdot\overline{\overline{I}} = \overline{\overline{I}}\cdot\mathbf{a} = \mathbf{a}$.

Dyadics for which $\overline{\overline{A}}^T = -\overline{\overline{A}}$ are called *antisymmetric* dyadics. Arbitrary dyadic can be uniquely decomposed into symmetric and antisymmetric parts:

$$\overline{\overline{A}} = \frac{1}{2}(\overline{\overline{A}} + \overline{\overline{A}}^T) + \frac{1}{2}(\overline{\overline{A}} - \overline{\overline{A}}^T) = \overline{\overline{A}}_s + \overline{\overline{A}}_a \tag{28}$$

Furthermore, any antisymmetric dyadic can be written as a vector product of a vector and the unit dyadic:

$$\overline{\overline{A}}_a = \mathbf{v}\times\overline{\overline{I}} \tag{29}$$

You might note that the basic theory of dyadics develops in parallel with the tensor or matrix algebra.

References

[1] I.V. Lindell, *Methods for Electromagnetic Field Analysis,* Oxford: Clarendon Press, 1992. Second edition, 1995.

Appendix B: Reciprocity theorem

Suppose that there are two source currents $\mathbf{J}_1$ and $\mathbf{J}_2$ in a medium [described by complex tensor parameters $\overline{\overline{\varepsilon}}(\mathbf{r})$, $\overline{\overline{\mu}}(\mathbf{r})$] which generate electromagnetic fields $\mathbf{E}_1$, $\mathbf{H}_1$ and $\mathbf{E}_2$, $\mathbf{H}_2$, respectively. Because of the linearity of the Maxwell equations we can write equations for these two fields separately:

$$\nabla \times \mathbf{E}_1 = -j\omega\overline{\overline{\mu}} \cdot \mathbf{H}_1 \tag{30}$$

$$\nabla \times \mathbf{H}_1 = j\omega\overline{\overline{\varepsilon}} \cdot \mathbf{E}_1 + \mathbf{J}_1 \tag{31}$$

$$\nabla \times \mathbf{E}_2 = -j\omega\overline{\overline{\mu}} \cdot \mathbf{H}_2 \tag{32}$$

$$\nabla \times \mathbf{H}_2 = j\omega\overline{\overline{\varepsilon}} \cdot \mathbf{E}_2 + \mathbf{J}_2 \tag{33}$$

Let us multiply the equations in the first set by the field vectors from the second set:

$$\mathbf{H}_2 \cdot \nabla \times \mathbf{E}_1 = -j\omega\mathbf{H}_2 \cdot \overline{\overline{\mu}} \cdot \mathbf{H}_1 \tag{34}$$

$$\mathbf{E}_2 \cdot \nabla \times \mathbf{H}_1 = j\omega\mathbf{E}_2 \cdot \overline{\overline{\varepsilon}} \cdot \mathbf{E}_1 + \mathbf{E}_2 \cdot \mathbf{J}_1 \tag{35}$$

Similarly,

$$\mathbf{H}_1 \cdot \nabla \times \mathbf{E}_2 = -j\omega\mathbf{H}_1 \cdot \overline{\overline{\mu}} \cdot \mathbf{H}_2 \tag{36}$$

$$\mathbf{E}_1 \cdot \nabla \times \mathbf{H}_2 = j\omega\mathbf{E}_1 \cdot \overline{\overline{\varepsilon}} \cdot \mathbf{E}_2 + \mathbf{E}_1 \cdot \mathbf{J}_2 \tag{37}$$

Next, we subtract (37) from (35):

$$\mathbf{E}_2 \cdot \nabla \times \mathbf{H}_1 - \mathbf{E}_1 \cdot \nabla \times \mathbf{H}_2 = j\omega(\mathbf{E}_2 \cdot \overline{\overline{\varepsilon}} \cdot \mathbf{E}_1 - \mathbf{E}_1 \cdot \overline{\overline{\varepsilon}} \cdot \mathbf{E}_2) + \mathbf{E}_2 \cdot \mathbf{J}_1 - \mathbf{E}_1 \cdot \mathbf{J}_2 \tag{38}$$

An important step here: If $\overline{\overline{\varepsilon}}$ is a symmetric matrix, that is, it equals to its transpose, $\overline{\overline{\varepsilon}} = \overline{\overline{\varepsilon}}^T$, then $\mathbf{E}_2 \cdot \overline{\overline{\varepsilon}} \cdot \mathbf{E}_1 - \mathbf{E}_1 \cdot \overline{\overline{\varepsilon}} \cdot \mathbf{E}_2 = 0$ in the right-hand side. Consider symmetric matrices $\overline{\overline{\varepsilon}}$ and $\overline{\overline{\mu}}$. Subtracting (36) from (34) we get, using the symmetry of $\overline{\overline{\mu}}$,

$$\mathbf{H}_2 \cdot \nabla \times \mathbf{E}_1 - \mathbf{H}_1 \cdot \nabla \times \mathbf{E}_2 = 0 \tag{39}$$

Next, we sum up (38) and (39) which yields

$$\underline{\mathbf{E}_2 \cdot \nabla \times \mathbf{H}_1} - \mathbf{E}_1 \cdot \nabla \times \mathbf{H}_2 + \mathbf{H}_2 \cdot \nabla \times \mathbf{E}_1 - \underline{\mathbf{H}_1 \cdot \nabla \times \mathbf{E}_2} = \mathbf{E}_2 \cdot \mathbf{J}_1 - \mathbf{E}_1 \cdot \mathbf{J}_2 \tag{40}$$

Note that the underlined terms together give

$$\mathbf{E}_2 \cdot \nabla \times \mathbf{H}_1 - \mathbf{H}_1 \cdot \nabla \times \mathbf{E}_2 = \nabla \cdot (\mathbf{H}_1 \times \mathbf{E}_2) \tag{41}$$

Also, the other two terms combine as

$$\mathbf{H}_2 \cdot \nabla \times \mathbf{E}_1 - \mathbf{E}_1 \cdot \nabla \times \mathbf{H}_2 = \nabla \cdot (\mathbf{E}_1 \times \mathbf{H}_2) \tag{42}$$

Finally we integrate (40) over an arbitrary volume V bounded by surface S. The result is, after applying the Gauss theorem,

$$\int_S (\mathbf{H}_1 \times \mathbf{E}_2 + \mathbf{E}_1 \times \mathbf{H}_2) \cdot d\mathbf{S} = \int_V (\mathbf{E}_2 \cdot \mathbf{J}_1 - \mathbf{E}_1 \cdot \mathbf{J}_2)\, dV \tag{43}$$

This relation is called *the Lorentz lemma.*

Consider the limiting case when the integration volume is the whole space. Then, assuming even negligible losses which are always present, we conclude that the surface integral vanishes, and the result is *the reciprocity theorem*

$$\int_V (\mathbf{E}_2 \cdot \mathbf{J}_1 - \mathbf{E}_1 \cdot \mathbf{J}_2)\, dV = 0 \tag{44}$$

Note that the only condition for its validity (besides the Maxwell equations) is the symmetry of matrices $\overline{\overline{\varepsilon}}$ and $\overline{\overline{\mu}}$. Otherwise, the system can be lossy or lossless, homogeneous or inhomogeneous.

Appendix C: Description of Matlab programs

The modified Marcatili's and Goell'd methods for calculation of dispersion characteristics of rectangular open dielectric waveguides with uniaxial cores with the axis parallel to the waveguide axis are provided on disk. The algorithms, described in detail is Chapter 5, Sections 5.3.1 and 5.3.2, have been realized in Matlab programming language.

In the corresponding directories there are the files named

```
marcatili.m
Goell.m
```

that are the main files calling some functions. Comments inside these files explain how to modify the input data for specific waveguides and frequency ranges. Programs have been checked to be working with MatLab versions 5.3 and 6.5.

Index

FSC
www.fsc.org
MIX
Papier aus verantwortungsvollen Quellen
Paper from responsible sources
FSC® C105338

If you have any concerns about our products,
you can contact us on
ProductSafety@springernature.com

In case Publisher is established outside the EU,
the EU authorized representative is:
Springer Nature Customer Service Center GmbH
Europaplatz 3, 69115 Heidelberg, Germany

Printed by Libri Plureos GmbH
in Hamburg, Germany